METHODS IN MOLECULAR BIOLOGY

Series Editor
John M. Walker
School of Life and Medical Sciences
University of Hertfordshire
Hatfield, Hertfordshire, AL10 9AB, UK

For further volumes:
http://www.springer.com/series/7651

Autophagy in Differentiation and Tissue Maintenance

Methods and Protocols

Edited by

Kursad Turksen

Ottawa Hospital Research Institute, Ottawa, ON, Canada

Humana Press

Editor
Kursad Turksen
Ottawa Hospital Research Institute
Ottawa, ON, Canada

ISSN 1064-3745 ISSN 1940-6029 (electronic)
Methods in Molecular Biology
ISBN 978-1-4939-8747-4 ISBN 978-1-4939-8748-1 (eBook)
https://doi.org/10.1007/978-1-4939-8748-1

Library of Congress Control Number: 2018952924

This Humana Press imprint is published by the registered company Springer Science+Business Media, LLC part of Springer Nature.
The registered company address is: 233 Spring Street, New York, NY 10013, U.S.A.

Preface

Autophagy is a crucial cellular process that regulates many cellular functions. That autophagy is crucial is underscored by the diseases and conditions that can arise when autophagy goes awry. Among areas where a better understanding of autophagy is required is its role in the maintenance of stem cell subpopulations and regulation of differentiation. To aid in accelerating advances on these topics, I sought to compile state-of-the-art protocols from researchers actively working in this area. I am grateful to all of the contributors for their generosity in sharing their tested and reproducible protocols for this series.

I acknowledge and thank Dr. John Walker, Editor in Chief of the *Methods in Molecular Biology* series, for his ongoing support.

I am also very grateful to Patrick J. Marton, Executive Editor, *Methods in Molecular Biology*, for his support and encouragement during the course of this project.

A very special thank you goes to David C. Casey, Editor for the *Methods in Molecular Biology* series, for his tireless help and advice that contributed greatly to the completion of this volume.

Ottawa, ON, Canada *Kursad Turksen*

Contents

Contributors

JAVAD ALIZADEH • *Department of Human Anatomy and Cell Science, Max Rady College of Medicine, Rady Faculty of Health Sciences, University of Manitoba, Winnipeg, MB, Canada*

OLEG ANICHTCHIK • *Peninsula Schools of Medicine and Dentistry, Plymouth University, Plymouth, Devon, UK*

MANMOHAN S. BAJAJ • *Stem Cell Lab, National Centre for Cell Science, Pune, India*

SREEPARNA BANERJEE • *Department of Biological Sciences, Middle East Technical University, Ankara, Turkey*

CHANDRA S. BHOL • *Department of Life Science, National Institute of Technology Rourkela, Rourkela, Odisha, India*

SUJIT K. BHUTIA • *Department of Life Science, National Institute of Technology Rourkela, Rourkela, Odisha, India*

DANIELE BRUNO • *Department of Biotechnology and Life Sciences, University of Insubria, Varese, Italy*

ROBERT BUTTON • *Peninsula Schools of Medicine and Dentistry, Plymouth University, Plymouth, Devon, UK*

MORENA CASARTELLI • *Department of Biosciences, University of Milano, Milano, Italy*

ZHIYONG CHENG • *Department of Human Nutrition, Foods, and Exercise, Fralin Translational Obesity Research Center, College of Agriculture and Life Science, Blacksburg, VA, USA*

FATMA KAYA DAGISTANLI • *Medical Biology Department, Cerrahpasa Faculty of Medicine, Istanbul University, Istanbul, Turkey*

DURGESH N. DAS • *Department of Medicine, University of Texas Health Science Center, Tyler, TX, USA*

EFTEKHAR EFTEKHARPOUR • *Department of Physiology and Pathophysiology, Regenerative Medicine, Program and Spinal Cord research Center, Max Rady College of Medicine, Rady Faculty of Health Sciences, University of Manitoba, Winnipeg, MB, Canada*

TRACEY EVANS • *Peninsula Schools of Medicine and Dentistry, Plymouth University, Plymouth, Devon, UK*

DU FENG • *Key Laboratory of Protein Modification and Degradation, School of Basic Medical Sciences, Affiliated Cancer Hospital and Institute of Guangzhou Medical University, Guangzhou Medical University, Guangzhou, People's Republic of China*

HUI GAO • *State Key Laboratory of Stem Cell and Reproductive Biology, Institute of Zoology, Chinese Academy of Sciences, Beijing, Chaoyang, People's Republic of China; University of Chinese Academy of Sciences, Beijing, People's Republic of China*

SAEID GHAVAMI • *Department of Human Anatomy and Cell Science, Max Rady College of Medicine, Rady Faculty of Health Sciences, University of Manitoba, Winnipeg, MB, Canada; Biology of Breathing Theme, Children's Hospital Research Institute of Manitoba, University of Manitoba, Winnipeg, MB, Canada; Health Policy Research Centre, Shiraz University of Medical Sciences, Shiraz, Iran*

DEVRIM GOZUACIK • *Faculty of Engineering and Natural Sciences, Molecular Biology, Genetics and Bioengineering Program, Sabanci University, Istanbul, Turkey; Center of*

Excellence for Functional Surfaces and Interfaces for Nano Diagnostics (EFSUN), Sabanci University, Istanbul, Turkey

ANNALISA GRIMALDI • *Department of Biotechnology and Life Sciences, University of Insubria, Varese, Italy*

KRISTYN GUMPPER • *Davis Heart and Lung Research Institute, The Ohio State University, Columbus, OH, USA; Department of Surgery, The Ohio State University Wexner Medical Center, Columbus, OH, USA*

MOHAMMAD HASHEMI • *Department of Clinical Biochemistry, Zahedan University of Medical Sciences, Zahedan, Iran*

RONGRONG HE • *Anti-Stress and Health Research Center, College of Pharmacy, Jinan University, Guangzhou, People's Republic of China*

YONGQUAN HU • *Key Laboratory of Protein Modification and Degradation, School of Basic Medical Sciences, Affiliated Cancer Hospital and Institute of Guangzhou Medical University, Guangzhou Medical University, Guangzhou, People's Republic of China*

ANNE-CLAIRE JACOMIN • *School of Life Sciences, University of Warwick, Coventry, UK*

ALON KAHANA • *Department of Ophthalmology and Visual Sciences, Kellogg Eye Center, University of Michigan, Ann Arbor, MI, USA*

VAIJAYANTI P. KALE • *Symbiosis Center for Stem Cell Research, Symbiosis School of Biological Sciences, Symbiosis International (Deemed University, Pune, India); Stem Cell Lab, National Centre for Cell Science, Pune, India*

PHILLIP E. KISH • *Department of Ophthalmology and Visual Sciences, Kellogg Eye Center, University of Michigan, Ann Arbor, MI, USA*

MUHAMMED KOCAK • *Faculty of Engineering and Natural Sciences, Molecular Biology, Genetics and Bioengineering Program, Sabanci University, Istanbul, Turkey*

GEORGIOS KONSTANTINIDIS • *Chemical Genomics Centre of the Max Planck Society, Dortmund, Germany; Max Planck Institute of Molecular Physiology, Dortmund, Germany*

ROHAN S. KULKARNI • *Stem Cell Lab, National Centre for Cell Science, Pune, India*

SHUPENG LI • *Institute of Neurology, Guangdong Key Laboratory of Age-Related Cardiac-Cerebral Vascular Disease, Affiliated Hospital of Guangdong Medical College, Zhanjiang, Guangdong, People's Republic of China*

WEI LI • *State Key Laboratory of Stem Cell and Reproductive Biology, Institute of Zoology, Chinese Academy of Sciences, Beijing, Chaoyang, People's Republic of China; University of Chinese Academy of Sciences, Beijing, People's Republic of China*

WEN LI • *Key Laboratory of Protein Modification and Degradation, School of Basic Medical Sciences, Affiliated Cancer Hospital and Institute of Guangzhou Medical University, Guangzhou Medical University, Guangzhou, People's Republic of China; Anti-Stress and Health Research Center, College of Pharmacy, Jinan University, Guangzhou, People's Republic of China*

YIFANG LI • *Anti-Stress and Health Research Center, College of Pharmacy, Jinan University, Guangzhou, People's Republic of China*

PEI-HUI LIN • *Davis Heart and Lung Research Institute, The Ohio State University, Columbus, OH, USA; Department of Surgery, The Ohio State University Wexner Medical Center, Columbus, OH, USA*

XIAOYING LIN • *Institute of Neurology, Guangdong Key Laboratory of Age-Related Cardiac-Cerebral Vascular Disease, Affiliated Hospital of Guangdong Medical College, Zhanjiang, Guangdong, People's Republic of China*

CHAO LIU • *State Key Laboratory of Stem Cell and Reproductive Biology, Institute of Zoology, Chinese Academy of Sciences, Beijing, Chaoyang, People's Republic of China; University of Chinese Academy of Sciences, Beijing, People's Republic of China*
LONGHUA LIU • *Department of Human Nutrition, Foods, and Exercise, Fralin Translational Obesity Research Center, College of Agriculture and Life Science, Blacksburg, VA, USA; Department of Pathology and Cell Biology, Naomi Berrie Diabetes Center, College of Physicians and Surgeons, Columbia University, New York, NY, USA*
SHOUQING LUO • *Peninsula Schools of Medicine and Dentistry, Plymouth University, Plymouth, Devon, UK*
KEWAL K. MAHAPATRA • *Department of Life Science, National Institute of Technology Rourkela, Rourkela, Odisha, India*
TIAN MENG • *Key Laboratory of Protein Modification and Degradation, School of Basic Medical Sciences, Affiliated Cancer Hospital and Institute of Guangzhou Medical University, Guangzhou Medical University, Guangzhou, People's Republic of China*
AURORA MONTALI • *Department of Biotechnology and Life Sciences, University of Insubria, Varese, Italy*
SUBHADIP MUKHOPADHYAY • *Department of Radiation Oncology, NYU Langone Medical Center, New York, NY, USA*
PRAJNA P. NAIK • *PG Department of Zoology, Vikram Deb (Auto) College, Jeypore, Odisha, India*
IOANNIS P. NEZIS • *School of Life Sciences, University of Warwick, Coventry, UK*
KIRAN SUHAS NILANGEKAR • *Developmental Biology Group, MACS-Agharkar Research Institute, Pune, India*
OZLEM DAGDEVIREN OZSOYLEMEZ • *Department of Radiobiology, Institute of Graduate Studies in Science and Engineering, Istanbul University, Istanbul, Turkey*
DENIZ GULFEM OZTURK • *Faculty of Engineering and Natural Sciences, Molecular Biology, Genetics and Bioengineering Program, Sabanci University, Istanbul, Turkey*
MELEK OZTURK • *Medical Biology Department, Cerrahpasa Faculty of Medicine, Istanbul University, Istanbul, Turkey*
PRASHANTA K. PANDA • *Institute of Cancer and Genomic Sciences, College of Medical and Dental Sciences, University of Birmingham, Birmingham, UK*
DEBASNA P. PANIGRAHI • *Department of Life Science, National Institute of Technology Rourkela, Rourkela, Odisha, India*
SRIMANTA PATRA • *Department of Life Science, National Institute of Technology Rourkela, Rourkela, Odisha, India*
PRAKASH P. PRAHARAJ • *Department of Life Science, National Institute of Technology Rourkela, Rourkela, Odisha, India*
ALFONSO SAERA-VILA • *Department of Ophthalmology and Visual Sciences, Kellogg Eye Center, University of Michigan, Ann Arbor, MI, USA*
SARBARI SAHA • *Department of Life Science, National Institute of Technology Rourkela, Rourkela, Odisha, India*
ADEL SEPANJNIA • *Department of Immunology, School of Medicine, Jiroft University of Medical Science, Jiroft, Iran*
MATTHEW SERMERSHEIM • *Davis Heart and Lung Research Institute, The Ohio State University, Columbus, OH, USA; Department of Surgery, The Ohio State University Wexner Medical Center, Columbus, OH, USA*

SHAHLA SHOJAEI • *Department of Human Anatomy and Cell Science, Max Rady College of Medicine, Rady Faculty of Health Sciences, University of Manitoba, Winnipeg, MB, Canada*
BHUPENDRA V. SHRAVAGE • *Developmental Biology Group, MACS-Agharkar Research Institute, Pune, India*
SONJA SIEVERS • *Max Planck Institute of Molecular Physiology, Dortmund, Germany; Compound Management and Screening Center of the Max Planck Society, Dortmund, Germany*
NIHARIKA SINHA • *Department of Animal Science, Michigan State University, East Lansing, MI, USA*
WON-HEE SONG • *Division of Animal Sciences, University of Missouri, Columbia, MO, USA; Department of Animal Science and Technology, Chung-Ang University, Anseong, South Korea*
PETER SUTOVSKY • *Division of Animal Sciences, University of Missouri, Columbia, MO, USA; Department of Obstetrics, Gynecology and Women's Health, University of Missouri, Columbia, MO, USA*
ZHIPENG TAO • *Department of Human Nutrition, Foods, and Exercise, Fralin Translational Obesity Research Center, College of Agriculture and Life Science, Blacksburg, VA, USA*
GIANLUCA TETTAMANTI • *Department of Biotechnology and Life Sciences, University of Insubria, Varese, Italy*
SINEM TUNÇER • *Department of Biological Sciences, Middle East Technical University, Ankara, Turkey*
BAOJIN WU • *Guangdong Landau Biotechnology Limited Company, Guangzhou, People's Republic of China*
YAO-WEN WU • *Chemical Genomics Centre of the Max Planck Society, Dortmund, Germany; Max Planck Institute of Molecular Physiology, Dortmund, Germany; Department of Chemistry, Umeå University, Umeå, Sweden*
LOUISE D. ZHENG • *Department of Human Nutrition, Foods, and Exercise, Fralin Translational Obesity Research Center, College of Agriculture and Life Science, Blacksburg, VA, USA*
MICHAEL X. ZHU • *Department of Integrative Biology and Pharmacology, McGovern Medical School, The University of Texas Health Science Center at Houston, Houston, TX, USA*

Methods in Molecular Biology (2018) 1854: 1–12
DOI 10.1007/7651_2018_168

Published online: 24 June 2018

Visualization and Measurement of Multiple Components of the Autophagy Flux

Tracey Evans, Robert Button, Oleg Anichtchik, and Shouqing Luo

Abstract

Autophagy is an intracellular degradation process that mediates the clearance of cytoplasmic components. As well as being an important function for cellular homeostasis, autophagy also promotes the removal of aberrant protein accumulations, such as those seen in conditions like neurodegeneration. The dynamic nature of autophagy requires precise methods to examine the process at multiple stages. The protocols described herein enable the dissection of the complete autophagy process (the "autophagy flux"). These allow for the elucidation of which stages of autophagy may be altered in response to various diseases and treatments.

Keywords Autolysosome, Autophagosome, Autophagy, Lysosomal integrated membrane protein (LIMP), Lysosome, Lysosome associated membrane protein (LAMP), Microtubule-associated protein 1A/B-light chain 3 (LC3), Sequestosome-1/p62

1 Introduction

Autophagy is an intracellular degradation process which mediates the clearance of various cytoplasmic materials, such as damaged organelles or aberrant protein aggregates [1]. Given that many of the processes' targets are associated with disease, autophagy has become an area of interest across a diverse range of studies. Therefore, reliable and accurate methods to assess autophagy activity are important in the pursuit for potential future breakthroughs.

In essence, autophagy proceeds through the capture of cytoplasmic portions and target material into double-membraned vesicles—autophagosomes [2]. These are then shuttled to the lysosomes, whereupon the two vesicles fuse. Lysosomal enzymes degrade the autophagosomes and their contents, with some of the resulting products recycled by the cell (summarized in Fig. 1) [3].

Owing to their crucial role in the process, autophagosome numbers are typically used as a measure of autophagy activity. However, this readout alone can be misleading; autophagy is a

Tracey Evans and Robert Button contributed equally to this work.

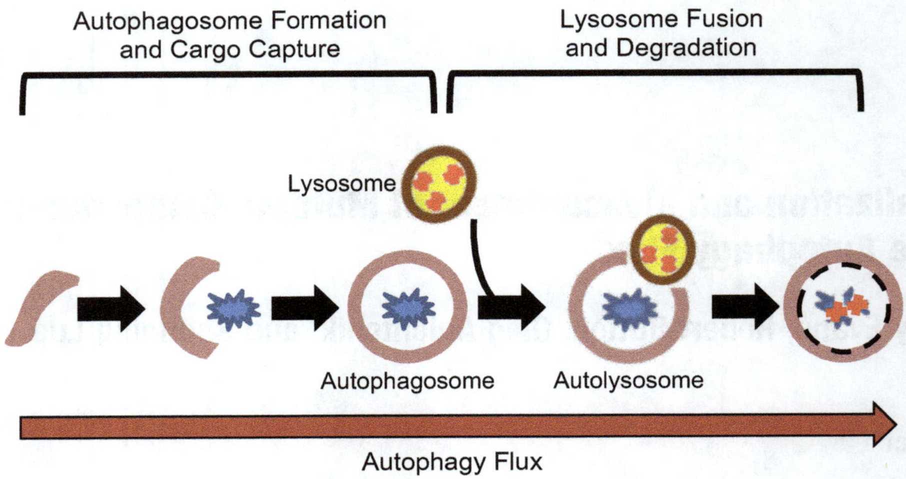

Fig. 1 A brief overview of autophagy. Autophagy begins with the formation of the double-membraned autophagosome, with cytoplasmic material captured in the process. Autophagosomes are then shuttled to the lysosomes, whereupon the two vesicles fuse, yielding the autolysosome. Lysosomal enzymes then degrade the autophagosome inner membrane and its cargo. The completion of this process is referred to as the "Autophagy Flux"

dynamic process, meaning appropriate cautions must be employed when making static measurements. For instance, while an increase in autophagosomes could represent enhanced autophagy, it could alternatively be the very different scenario of a failure in the lysosomal clearance of the vesicles [4]. Therefore, attention to the complete process—the "autophagy flux"—is the most accurate approach to take when studying autophagy, and will be the focus of this chapter.

2 Materials

Unless stated, reagents and chemicals were from ThermoFisher (UK).

1. *Positive controls for autophagy flux inhibition*:
   Bafilomycin-A1 (Millipore, 19-148) and Chloroquine (Sigma Aldrich, C6628) both stored at −20 °C and soluble in DMSO (dimethyl sulfoxide).
2. *Lysosome dyes*:
   Acridine Orange (ThermoFisher, A1301)—prior to use, dilute to 2.5 mg/ml in dH_2O.
   The LysoSensor range of dyes are all available from ThermoFisher:
   LysoSensor Blue DND-167 (#L7533)
   LysoSensor Green DND-189 (#L7535)

LysoSensor Yellow/Blue DND-160 (#L7545)
LysoSensor Yellow/Blue 10,000 MW dextran (#L22460)

3. *Double Immunofluorescence assay*:
Microscope slides vessel: Glass coplin jars are used for up to ten microscope slides, a larger vessel is required for more slides.
Deparaffinization: Xylene (Sigma Aldrich, UK).
Tissue rehydration: Ethanol diluted to required concentrations with distilled H_2O.
Tissue barrier pen: PAP pen (liquid blocker, Sigma Aldrich, UK).
Autofluorescence quenching: 0.1% Sudan black (Sigma Aldrich, UK), dissolved in 70% ethanol.
Antigen retrieval: 10 mM Citrate buffer dissolved in water (pH 6.0).
5% Normal goat serum (NGS) is diluted in PBS with 0.05% Triton-X100 (PBS-T).
Primary antibodies: Beta-synuclein (1:250; Abcam, Ab76111) and LIMP2 (1:500; ThermoFisher PA5-20540).
Secondary antibodies: Alexa Fluor conjugated secondary antibodies (1:1000; Invitrogen Molecular Probes) diluted in PBS-T with
Hoechst nuclei stain (1:1000; New England Biolabs, UK, 33342).
Mounting coverslips on tissue sections: PermaFluor Aqueous Mounting Medium (ThermoFisher, TA006-FM).

3 Methods

3.1 Autophagosome Accumulation

The classic marker for studying autophagosomes is microtubule-associated protein 1A/B-light chain 3 (LC3)-II. LC3 is cleaved at its C-terminus by Atg4 to form LC3-I, and LC3-I is covalently conjugated to PE to form LC3-II. LC3-II remains associated with autophagosome membranes through to degradation. Therefore, LC3-II levels can be measured with a variety of techniques, perhaps most commonly with immunoblotting or immunocytochemistry. However, purely measuring LC3-II levels or counting LC3-II puncta is of limited use owing to the dynamic nature of autophagy.

Interpreting what changes to LC3-II mean for autophagy flux requires some careful consideration. While an increase in LC3-II levels or autophagosome numbers may appear to indicate increased autophagy activity, it could instead be due to a reduction in their degradation, resulting in their accumulation. To address this point of contention, treatment with a positive control for autophagy flux inhibition can be used. Treatments with lysosome acidity inhibitors like Bafilomycin-A1 (100–400 nM, 4 h) and Chloroquine (25 μM,

Table 1
Assessing autophagy flux changes via LC3 analysis

Observation	Autophagy activity
Decreased LC3-II levels/puncta	Decreased autophagosome synthesis
Increased LC3-II levels/puncta	Increased autophagosome synthesis OR block to autophagy flux
Further increase with lysosome inhibitor	Increased autophagosome synthesis
No increase with lysosome inhibitor	Decreased autophagy flux

8–16 h) will block the autophagy flux and lead to autophagosome accumulation. These can then be used for comparison with any experimental treatments of your own. If the levels/numbers of LC3 positive vesicles increase further when your treatment is combined with Bafilomycin-A1/Chloroquine, this suggests that your treatment enhances autophagosome synthesis. However, if combining your treatment with Bafilomycin-A1/Chloroquine causes no change, this indicates that it already blocks the autophagy flux.

For ease of reference, this is summarized in Table 1.

3.2 mRFP-GFP-LC3 Stably Expressing "Traffic Light" Cells

HeLa cells stably expressing mRFP-GFP-LC3 (sometimes referred to as "Traffic Light" cells) provide a system to distinguish between autophagosomes (non-degraded) and autolysosomes (autophagosomes fusing with lysosomes). This is down to the differing pH sensitivities of mRFP and GFP: GFP is quenched at an acidic pH more readily than mRFP. Therefore, any autophagosomes that fuse with lysosomes will lose GFP signal while retaining mRFP [5]. Alternative cell types can be used for this type of assay by transient transfection with mRFP-GFP-LC3 plasmid DNA.

1. Maintain cells in standard DMEM (10% FBS, 1% penicillin/streptomycin) supplemented with 100 μg/ml G418 as a selection agent. Cells are cultured at 37 °C, with 5% CO_2.
2. Plate at appropriate density and treat according to requirements. For recommended positive controls, *see* additional **Note 1**.
3. Remove media and rinse with 1× PBS before fixing cells with 4% paraformaldehyde (PFA) for 10 min.
4. Wash three times with 1× PBS.
5. If additional staining is required, proceed with usual antibody labeling protocols. *See* additional **Note 2**.
6. Mount coverslips and take for microscopy.
7. Image analysis. *See* additional **Note 3**.

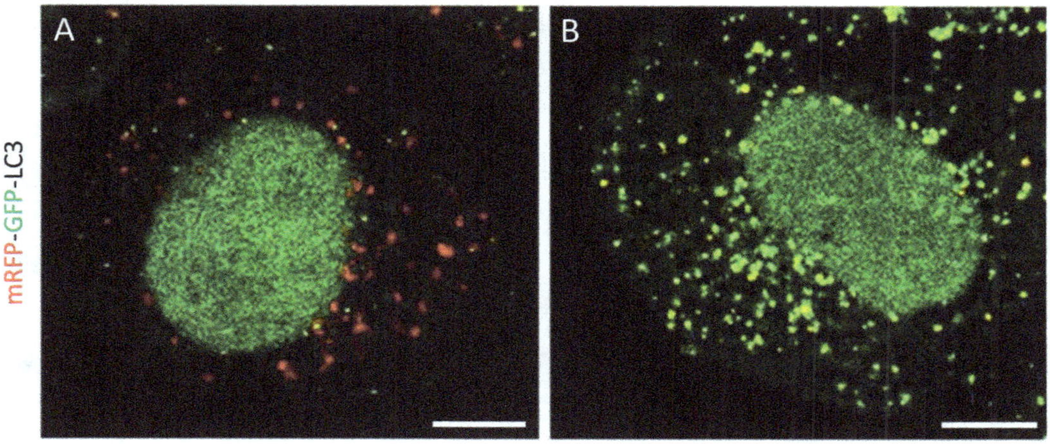

Fig. 2 "Traffic Light" (mRFP-GFP-LC3) cells treated with (**a**) vehicle or (**b**) lysosome inhibitor (Bafilomycin-A1). Note the increase in yellow (non-degraded) autophagosome puncta when autophagy flux is inhibited. Scale bar is 10 μm

Autophagosomes	All GFP-positive vesicles
Autolysosomes	Number of RFP vesicles minus the number of GFP vesicles

[a]Conditions in which autophagy flux is blocked will result in an increase in autophagosome numbers and a decrease in autolysosomes. Visually, this will be clear by elevated numbers of yellow vesicles (Fig. 2).

3.3 p62: The Autophagy Substrate

Another indicator of the autophagy flux is the autophagy substrate p62 (also known as Sequestosome-1). p62 serves as an adaptor to link autophagy targets to the autophagosomes, allowing for the selective capture and removal of material. p62 is engulfed and taken for lysosomal degradation in the process [6]. Therefore, levels of p62 are linked to the state of autophagy flux.

Assays for p62 can be performed with both immunoblotting and imaging techniques, measuring band densities in the former and number of cytoplasmic puncta in the latter (*see* Fig. 3).

As p62 is an autophagy substrate, it follows that its levels should have an inverse correlation with the extent of autophagy activity. Indeed, a decrease in p62 would suggest an increased rate of degradation, and thus elevated autophagy. However, if p62 levels increase, this could mean one of two different scenarios: (1) there could be a decrease in autophagosome synthesis, or (2) autophagy flux could be inhibited. This uncertainty can be resolved with the use of lysosome inhibitors like Bafilomycin-A1 and Chloroquine. In Option (1), p62 levels will increase further with flux blockage, whereas in Option (2) there will be little change, since degradation is already defective.

For ease of reference, this is summarized in Table 2.

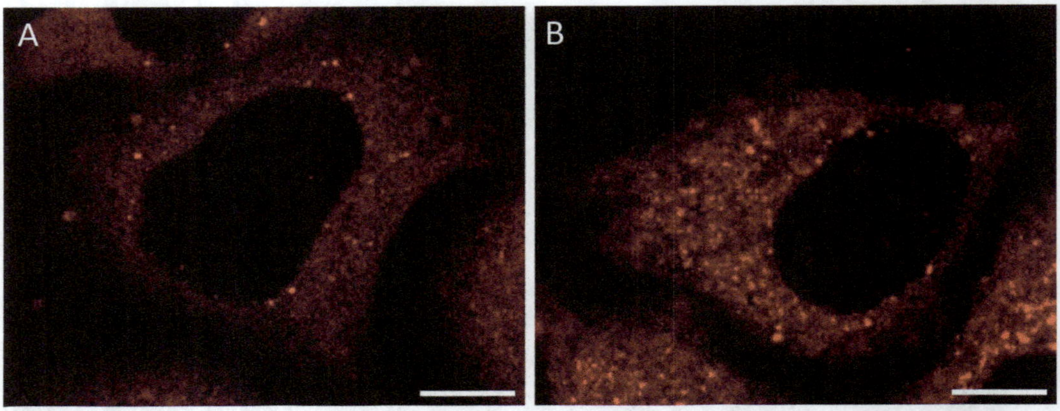

Fig. 3 HeLa cells stained with p62 antibody (1:1000, BD Biosciences, cat # 610833). Cells are treated with either vehicle (**a**) or chloroquine for 16 h (**b**). Note the increase in p62 puncta when lysosomal degradation is inhibited. Scale bar is 10 μm

Table 2
Assessing autophagy flux changes via p62 analysis

Observation	Autophagy activity
Decreased p62 levels/puncta	Increased autophagy/autophagy flux (as enhanced p62 degradation)
Increased p62 levels/puncta	Decreased autophagosome synthesis OR block to autophagy flux
Further increase with lysosome inhibitor	Decreased autophagosome synthesis
No increase with lysosome inhibitor	Decreased autophagy flux

3.4 Lysosomes

Effective lysosomal function is crucial for a healthy autophagy flux. A drop in the lysosome's acidic pH or damage to its structural integrity will prevent the degradation of autophagosomes, therefore halting the completion of autophagy. This means assessment of lysosomal conditions can provide a valuable insight into the state of the end stage of autophagy. We provide two protocols for exploring lysosomal acidity: Acridine Orange and LysoSensor.

Note that both Acridine Orange and LysoSensor staining require cells to still be alive, hence *no fixation is necessary.*

3.4.1 Acridine Orange

Acridine Orange (AO) is a lysosomotropic metachromatic fluorochrome. It can be used as a measure of intracellular acidity. While Acridine Orange usually expresses green fluorescence, it can enter lysosomes whereupon it is protonated and shifts to red ("orange") fluorescence [7, 8]. Therefore, this provides a means to differentiate between acidic and nonacidic compartments in cells.

1. Plate cells at appropriate density, and treat as desired. *See* additional **Note 4** for recommended positive controls.
2. Prepare Acridine Orange solution at 2.5 μg/ml in cell culture media.
3. The total volume needed must be sufficient to cover all cells to be tested. *See* **Note 5** for guide volumes.
4. Remove media from cells, wash with 1× PBS, and replace with Acridine Orange solution.
5. Incubate cells for 15 min at 37 °C.
6. Wash cells twice with 1× PBS.
7. Replace with fresh media.
8. Take for live cell imaging. Use an excitation wavelength of 488 nm; collect emission at both 505–570 nm and 615–754 nm. Acidic vesicles (lysosomes) should appear orange/red; nonacidic regions green.
9. Analysis:

   Control cells should show a distribution of small orange/red vesicles scattered in the green cytoplasm. A decrease or loss of these vesicles is indicative of loss of acidity, which would suggest a blockade to autophagy flux. Other alterations such as changes to lysosome size and distribution may also result in autophagy flux disruption. Example cells are shown in Fig. 4.

3.4.2 LysoSensor

The LysoSensor series of probes are acidotropic agents that, like Acridine Orange, can accumulate in lysosomes and show a shift in fluorescence upon protonation. LysoSensor comes with a choice of colors (Blue, Green, and Yellow/Blue), but generally all show enhanced fluorescence upon protonation.

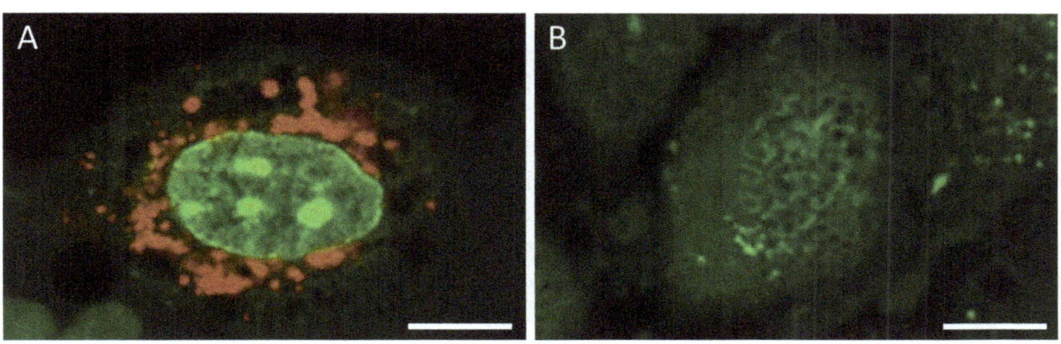

Fig. 4 HeLa cells stained with Acridine Orange, treated with either vehicle (**a**) or lysosomal deacidifier Bafilomycin-A1 (400 nM) (**b**) for 4 h. Note the red acidic vesicles (lysosomes), which are lost upon Bafilomycin-A1 treatment. Scale bar is 10 μm

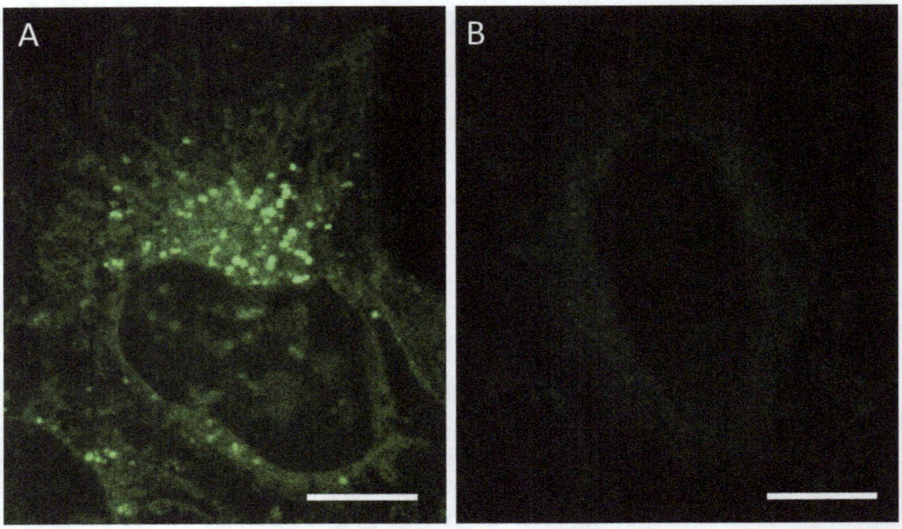

Fig. 5 HeLa cells stained with LysoSensor Green DND-189, treated with either vehicle (**a**) or Bafilomycin-A1 (400 nM) (**b**) for 4 h. Note the fluorescent vesicles (lysosomes) under control conditions, which are lost upon Bafilomycin-A1 treatment. Scale bar is 10 μm

Protocol

1. Plate cells at appropriate density, and treat as desired. *See* additional **Note 6** for positive controls.
2. Stain cells with 1 μM LysoSensor for 30 min at 37 °C. Remove media and wash cells twice with 1× PBS.
3. Replace with fresh media.
4. Perform live cell imaging. *See* additional **Note 7**.
5. Analysis. *See* additional **Notes 8** and **9** and Fig. 5.

As autophagy flux is implicated in numerous diseases, these protocols can be applied to multiple fields. To illustrate this, we have included a "practical" example of assessing autophagy flux in a model of neurodegeneration below.

3.5 Visualizing the Autophagy Flux in Postmortem Brain Tissues from Dementia Patients

The neurodegenerative process can occur in parallel with dysfunction to protein clearance pathways that would usually serve to disrupt the aggregating process. Such dysfunction is observed in pathologies such as dementia with Lewy bodies (DLB) [9] and Alzheimer's disease [10]. In these cases, it is essential that protocols are developed that enable key autophagy markers and the offending protein aggregates to be observed in tandem, allowing us to elucidate the stage of the degradation pathway that is functionally impaired. Since the autophagy process can be disrupted at various points, use of markers associated with autophagosome formation (LC3-II), lysosomes (LAMP-2 and LIMP2), autophagosome–lysosome fusion, and

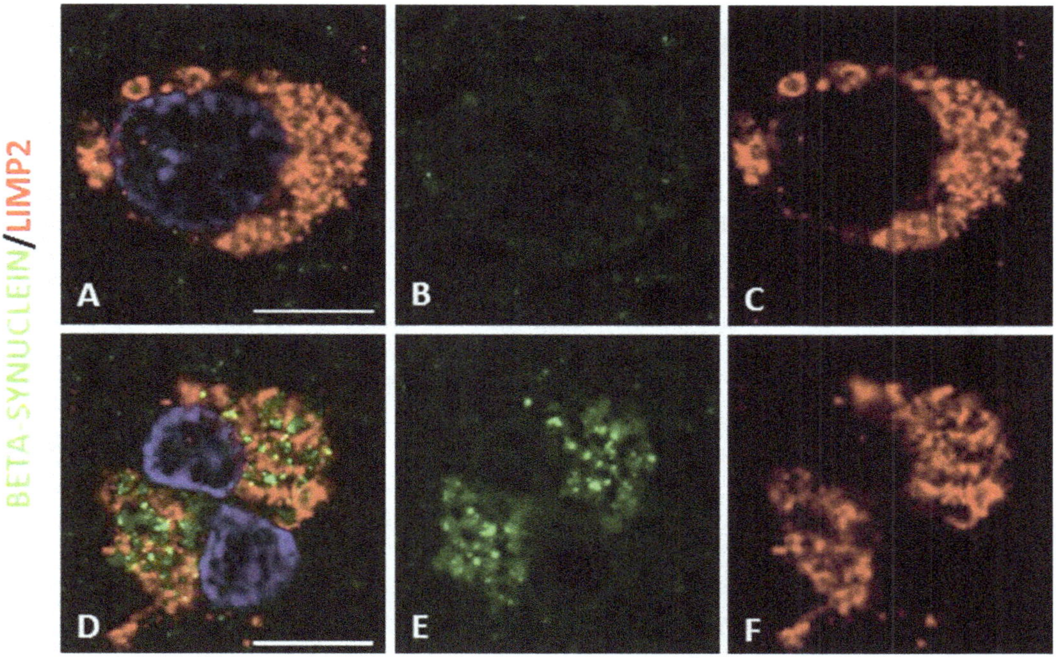

Fig. 6 Cortical changes in autophagy in the dementia with Lewy bodies (DLB) cortex (**d–f**) when compared to the control (**a–c**). The double immunofluorescent protocol shows a positive β-synuclein signal (1:250) (**e**) within distorted LIMP2-immunopositive structures (1:500) (**f**). Scale bar is 10 μm

substrate degradation (p62) will help to reveal which point in the process is compromised. To address this, a double immunofluorescence protocol has been optimized and employed, allowing the examination of the distribution pattern and cellular location of autophagy-related markers in relation to the presence of proteins involved in neurodegeneration [9]. This method enabled the examination of lysosomal marker LIMP2 in relation to the presence of the β-synuclein protein in DLB (Fig. 6). LIMP2/SCARB2 is associated with lysosomal pathology in α-synucleinopathies [11].

Since the burden of pathological aggregates appears to drive the neurodegenerative phenotype, methods such as the one we outline here may be of great importance in elucidating and manipulating processes to oppose degenerative progression.

Protocol for Double Immunofluorescent Immunohistochemistry

1. Formalin fixed and paraffin embedded brain tissue sections (6 μm) are deparaffinized in two changes of xylene for 5 min each. The toxic nature of xylene requires this deparaffinization to be performed in a hood according to the manufacturer's instructions.

2. Subsequent tissue rehydration is performed in a stepwise serial dilution of ethanol for 2 min each (100%, 95%, 80%, and 70%, diluted in distilled H_2O).
3. Tissue sections are washed briefly in H_2O before being equilibrated in 1× PBS for 5 min.
4. Deposits such as lipofuscin are particularly prominent in aging brains. In order to quench autofluorescence produced by such deposits, tissue incubation in 0.1% Sudan black, diluted in 70% ethanol, is carried out for 10 min at room temperature followed by washing in PBS with 0.05% Triton X100 (1×PBS-T). *See* additional **Note 10** for more detail related to Sudan black.
5. Antigen retrieval will be dependent on the protein of interest. Our protocol involves boiling the tissue in 10 mM of citrate buffer (pH 6.0) in a microwave oven for 30 min at 700 W. *See* additional **Note 11**. Microscope slides are placed into the citrate buffer in a glass staining rack holder, the citrate buffer is held in a microwaveable plastic container with a vent for steam release during boiling.
6. Following boiling in citrate buffer, tissue sections are washed in running water for 10 min.
7. Using a PAP pen, mark the area surrounding the tissue of interest and proceed to the blocking step.
8. Blocking requires tissue incubation with appropriate corresponding normal serum (5%) in 1× PBS-T for 1 h at room temperature. The normal serum will be the species in which the secondary antibody has been raised.
9. Primary antibodies are diluted in PBS-T at the optimized concentration and incubated with the tissue overnight at +4 °C. (*See* Fig. 6 for primary antibody concentrations.)
10. Wash three times with 1× PBS before applying the Alexa-conjugated secondary antibodies diluted in 1× PBS-T, 5% normal blocking serum and Hoerscht (1:1000) for nuclear staining. Incubation occurs in the dark at room temperature for 2 h.
11. Following appropriate washing with 1× PBS, the slides are coverslipped with PermaFluor mounting medium in preparation for confocal laser scanning microscopy.

4 Notes

1. Recommended positive controls for autophagy flux inhibition: Lysosomal deacidifying drugs such as Bafilomycin-A1 (100 nM, 4 h) or Chloroquine (25 μM, 16 h) should lead to a block to autophagy flux.

2. Traffic Light cells already express GFP- and RFP-LC3; therefore, these channels cannot be used for any additional staining.
3. Vesicles positive for GFP AND RFP-LC3 (YELLOW): non-degraded autophagosomes; vesicles positive for RFP-LC3 ONLY: autolysosomes.
4. Recommended positive control for lysosomal deacidification: Lysosomal deacidifying drugs such as Bafilomycin-A1 (100 nM, 4 h) or Chloroquine (25 μM, 16 h) should lead to a decrease/loss of visible acidic vesicles.
5. Guide volumes per well: 1 ml (6-well plate); 0.5 ml (12-well plate), etc.
6. Recommended positive control for lysosomal deacidification: Lysosomal deacidifying drugs such as Bafilomycin-A1 (100 nM, 4 h) or Chloroquine (25 μM, 16 h) should lead to a decrease/loss of visible acidic vesicles.
7. Excitation wavelength will differ depending on which LysoSensor color has been used.
8. Control cells should exhibit brighter fluorescent vesicles (lysosomes) in the cytoplasm, with a loss of these in lysosome inhibitory treatments (*see* Fig. 5).
9. The LysoSensor Yellow/Blue strongly acidic compartments should appear yellow, with weaker regions blue. Treatments that reduce acidity should show a decline/loss in these vesicles.
10. The autofluorescence associated with fluorescent microscopy is particularly prevalent in the aging brain due to the presence of deposits such as lipofuscin. Sudan black dye is able to bind to deposits in the human brain and reduce the nonspecific signal produced. Variations in Sudan black concentration may be required during the optimization process.
11. Antigen retrieval methods will be dependent on the proteins of interest. Citrate buffer or TRIS–EDTA are commonly appropriated methods that involve boiling the tissue samples, and further methods including incubation in formic acid may be required for some proteins, for example, α-synuclein oligomers [9]; an aggregating protein involved in Parkinson's disease and Lewy body dementia.

References

1. Ravikumar B, Sarkar S, Davies JE, Futter M, Garcia-Arencibia M, Green-Thompson ZW et al (2010) Regulation of mammalian autophagy in physiology and pathophysiology. Physiol Rev 90(4):1383–1435
2. Klionsky DJ, Ohsumi Y (1999) Vacuolar import of proteins and organelles from the cytoplasm. Annu Rev Cell Dev Biol 15:1–32
3. Klionsky DJ (2005) The molecular machinery of autophagy: unanswered questions. J Cell Sci 118(Pt 1):7–18
4. Button RW, Roberts SL, Willis TL, Hanemann CO, Luo S (2017) Accumulation of autophagosomes confers cytotoxicity. J Biol Chem 292 (33):13599–13614

5. Kimura S, Noda T, Yoshimori T (2007) Dissection of the autophagosome maturation process by a novel reporter protein, tandem fluorescent-tagged LC3. Autophagy 3 (5):452–460
6. Bjorkoy G, Lamark T, Brech A, Outzen H, Perander M, Overvatn A et al (2005) p62/SQSTM1 forms protein aggregates degraded by autophagy and has a protective effect on huntingtin-induced cell death. J Cell Biol 171(4):603–614
7. Button RW, Vincent JH, Strang CJ, Luo S (2016) Dual PI-3 kinase/mTOR inhibition impairs autophagy flux and induces cell death independent of apoptosis and necroptosis. Oncotarget 7(5):5157–5175
8. Boya P, Kroemer G (2008) Lysosomal membrane permeabilization in cell death. Oncogene 27(50):6434–6451
9. Evans T, Kok WL, Cowan K, Hefford M, Anichtchik O (2018) Accumulation of beta-synuclein in cortical neurons is associated with autophagy attenuation in the brains of dementia with Lewy body patients. Brain Res 1681:1–13
10. Nixon RA, Wegiel J, Kumar A, Yu WH, Peterhoff C, Cataldo A et al (2005) Extensive involvement of autophagy in Alzheimer disease: an immuno-electron microscopy study. J Neuropathol Exp Neurol 64(2):113–122
11. Bras J, Guerreiro R, Darwent L, Parkkinen L, Ansorge O, Escott-Price V et al (2014) Genetic analysis implicates APOE, SNCA and suggests lysosomal dysfunction in the etiology of dementia with Lewy bodies. Hum Mol Genet 23(23):6139–6146

Methods in Molecular Biology (2018) 1854: 13–20
DOI 10.1007/7651_2018_167

Published online: 20 July 2018

Mitochondrial Redox Sensor for *Drosophila* Female Germline Stem Cells

Kiran Suhas Nilangekar and Bhupendra V. Shravage

Abstract

Mitochondrial reactive oxygen species (mROS), a major source of ROS within cells, functions as an important signaling molecule and has the ability to damage cellular macromolecules including DNA and proteins. Monitoring mROS levels is therefore essential to understand cell–cell communication and programmed cell death in all types of cell including stem cells. Here, we describe generation and characterization of a redox sensor for mROS that is specifically expressed in the germline stem cells (GSCs) in *Drosophila*. This redox sensor can be used to monitor the production of mROS and mitophagy in the GSCs during oogenesis.

Keywords *Drosophila*, Germline stem cells, Mitophagy, Mito-roGFP2-Grx1, Redox

1 Introduction

Autophagy is a degradative process that is crucial for maintaining protein and organelle homeostasis in distinct types of cells including stem cells. The damaged and defective mitochondria are degraded by the autophagic process termed as mitophagy [1, 2]. Mitochondria generate ATP which is essential for both anabolic and catabolic processes within the cells. Mitochondrial reactive oxygen species (mROS) is the by-product of these processes. Mitochondria can be damaged and become dysfunctional due to exposure to mROS. This impacts generation of ATP and further causes increased production of mROS which can lead to oxidative damage to macromolecules including DNA, RNA, and proteins [3, 4]. Thus, maintaining healthy mitochondria via the process of mitophagy is crucial for maintaining cellular homeostasis.

Incomplete reduction of oxygen during mitochondrial respiration leads to the formation of ROS. There are three primary species of ROS, i.e., superoxide ion ($O_2\bullet^-$), hydrogen peroxide (H_2O_2), and hydroxyl ion (OH^-). $O_2\bullet^-$ ion is produced when electrons are transferred from electron transfer chain complex I and complex III localized in the mitochondrial matrix. $O_2\bullet^-$ ions are reduced by superoxide dismutase, Cu–Zn-dependent SOD1 (localized in the mitochondrial intermembrane) and manganese-dependent SOD2 (localized to the mitochondrial matrix) to H_2O_2. Formation of OH^- ions from H_2O_2 can be catalyzed by nonenzymatic reaction mediated by Cu^+ or Fe^{2+}. Alternatively, H_2O_2 can be detoxified to H_2O by glutathione peroxidase (GPx) and peroxiredoxin III (PrxIII) [3–7]. The levels of $O_2{\bullet}^-$, H_2O_2, and OH^- are in a dynamic equilibrium mediated by antioxidant proteins and enzymes including glutaredoxins.

Glutaredoxins (Grx) are small redox enzymes that mediate reduction of substrates and in the process, themselves get oxidized [8, 9]. The redox state of glutaredoxins is maintained by nonenzymatic reduction by glutathione, a vital antioxidant protein synthesized within the cells. The ratio of reduced glutathione (GSH) to oxidized glutathione (GSSG) can provide useful information about the redox state of the cell and is frequently used as a measure of oxidative stress. The drawback of most methods to measure the ratio is that they disrupt the physical integrity of cells and are prone to introduction of artefacts. Also, such methods are unable to resolve the redox states spatially at subcellular or cellular levels. Redox-sensitive GFP (roGFP) has made it possible to measure redox in vivo without destroying the cells [10].

Recent advances in making fusion constructs with roGFP have led to resolution of redox states even at organelle level. Human glutaredoxin Grx1 fused to roGFP2 (Grx1-roGFP2) was first described by Marcus Gutscher and colleagues which was modified by adding a mitochondrial localization signal (mito-roGFP2-Grx1) by Albrecht et al. [8, 11]. We describe a transgenic line that expresses mito-roGFP2-Grx1 in the *Drosophila* germline under a germline-specific promoter that also functions in female germline stem cells (GSCs).

The mito-roGFP2-Grx1 cassette is cloned downstream of *nanos* promoter which allows for expression specifically in the GSCs and their progeny. The probe senses the redox state of Grx1 through roGFP2. Grx1 is oxidized nonenzymatically during conversion of oxidized glutathione (GSSG) to reduced glutathione (GSH). The redox state of Grx1 causes reduction or oxidation of the engineered cysteine residues of roGFP2 which are responsible for the switch in excitation of fluorescence (488–405 nm). The shift in excitation of roGFP2 is a reliable measure of the GSH redox states. The addition of mitochondrial localization signal at the 5′-terminus of *roGFP2-Grx1* leads to translocation of roGFP2-Grx1 protein to mitochondria, thus enabling measurement of redox potential within the mitochondria [11]. Once the dynamic range for redox state of the experiment is determined, the sensor can be used to know the in vivo extent of redox potential in the germline including GSCs.

2 Materials

1. Redox conservative reagent: 20 mM *N*-ethyl maleimide (NEM). Dissolve 25 mg of NEM in 10 ml of 1× PBS. Store at 4 °C. Use for max. one month from the date of preparation.
2. Oxidizing reagent: 4 mM diamide (DA). Dissolve 3.44 mg of DA in 5 ml of 1× PBS. Store at −20 °C.
3. Reducing reagent: 40 mM dithiothreitol (DTT). Dissolve 30.84 mg of DTT in 5 ml of 1× PBS. Store at −20 °C.
4. Fixative: 4% paraformaldehyde in 1× PBS (pH = 7.4).
5. Materials for dissection; Grace's medium, Dumont no. 5 forceps, no. 3 forceps, and minutien pins.

3 Methods

3.1 Generation of pC4 Nosp Mito-roGFP2-Grx1 Nos3′ UTR Transgenic Line

nanos3′UTR was PCR amplified from CantonS genome using primers nos3′UTRF25′-TCTAGAAGAGGGCGAATCCAGCTC TGGAGCAGA and nos3′UTRR 5′-TCTAGACCATTTTGGG AGACGCCTTGAACCTAAGTG and digested with XbaI and PstI. The resulting fragment of 1236 bp was cloned in XbaI, PstI digested pCasper4 to obtain pC4-nos3′UTR. Nanos promoter was amplified using primers nosPF 5′-AAGCTTCGACCGTTTT AACCTCGAAATATG and newnosPR 5′-GGCGAAAATCCGG GTCGAAAGTTACG to obtain 935-bp fragment. This fragment was cloned in pGEMt-Easy to obtain pGEMt-nosP. pGEMt-nosP was digested using EcoRI and the resulting 963-bp fragment was cloned in EcoRI digested pC4-nos3′UTR to obtain pC4-nosP-nos3′UTR. 1241-bp NotI-XbaI fragment from pUASt-mito-roGFP2-Grx1 was cloned in NotI-XbaI digested pC4-nosP-nos3′UTR to generate pC4-nosP-mito-roGFP2-Grx1-nos3′UTR. The entire cassette consisting of nosP-mito-roGFP2-Grx1-nos3′UTR was sequenced and confirmed (Fig. 1). Transgenics were generated at C-CAMP, Bangalore, India.

3.2 Dissection of Ovaries and In Vitro Treatment with Redox Reagents

In order to obtain the dynamic range of the redox biosensor, the tissue has to be fully oxidized and fully reduced. To avoid artificial oxidation of the tissue during fixation and further processing, the probe has to be chemically blocked before fixation using the redox conservative reagent.

1. Fatten ovaries of the transgenic flies by feeding them on yeast for at least 2 days. Dissect out ovaries from at least five flies (preferably ten flies). Perform the dissection in warm Grace's medium (25 °C) on a cavity slide. Tease apart the ovarioles using the minutien pins. Transfer the ovaries from the cavity

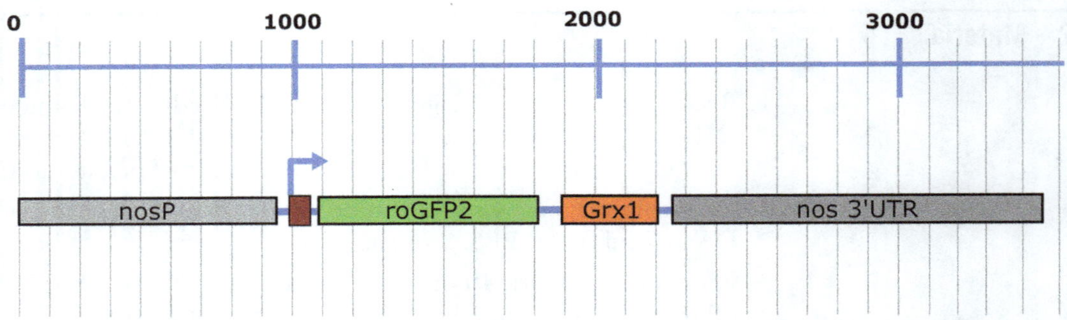

Fig. 1 Illustration showing the expression cassette nosP-mito-roGFP2-Grx1-nos3′UTR. Blue line depicts the size in base pairs (bp)

slide to a 1.5-ml tube using a wide bore 200 μl tip. To ensure that the ovaries do not stick to the tip, passivate the tip by loading it with 10% BSA solution for 2 min.

2. Let the ovaries settle to the bottom of the 1.5 ml tube. Remove the Grace's medium. Ensure not to aspirate the ovarioles. Wash the tissue by adding 500 μl of 1× PBS to the tube. Keep the tube on a 3D shaker at 20 rpm at room temperature for 2 min (*see* **Notes 1** and **2**).
3. For complete oxidation of the tissue, replace the 1× PBS with 500 μl of 4 mM DA solution and incubate for 10 min at room temperature with shaking at 5 rpm on a 3D shaker (*see* **Notes 3** and **4**).
4. For complete reduction of the tissue, replace the 1× PBS with 500 μl of 40 mM DTT solution and incubate for 10 min at room temperature with shaking at 5 rpm on 3D shaker (*see* **Notes 3** and **4**).
5. Wash the tissue once with 500 μl 1× PBS for 2 min after treatment with oxidizing or reducing reagent.
6. For chemically blocking the redox state of the tissue, it has to be treated with the redox conservative reagent. This state can be the above completely oxidized or reduced or the untreated test sample. Replace the 1× PBS with 500 μl of 20 mM NEM solution and incubate for 10 min at room temperature on a 3D shaker with shaking at 5 rpm.
7. Wash once with 500 μl 1× PBS for 2 min.
8. Fix the tissue with 500 μl of 4% paraformaldehyde for 15 min at room temperature (*see* **Note 5**).
9. Wash twice with 500 μl 1× PBS for 10 min each.
10. Replace the PBS with 80 μl of 80% glycerol. Transfer the ovarioles along with the glycerol from the tube to a clean glass slide using a wide bore tip. Separate the ovarioles

and egg chambers using the minutien pins. Place a coverslip. Make sure that there are no air bubbles trapped between the slide and the coverslip. Seal using transparent nail varnish.

3.3 Imaging of the Redox Biosensor Mito-roGFP2-Grx1

Excite the sample with 405 nm as well as 488 nm sequentially (line by line). Collect the emission between 500 and 530 nm for both excitations. For Figs. 2 and 3, imaging was done on the Leica TCS SP8 Confocal microscope with the following settings; pinhole: 2, emission: 500–530 nm, laser power: 20%, gain: 750, and frame accumulation: 6. Excitation was sequential at 405 and 488 nm (line by line). 16 bit images with 1024 × 1024 pixel resolution at 100 Hz scan speed (*see* **Note 6**).

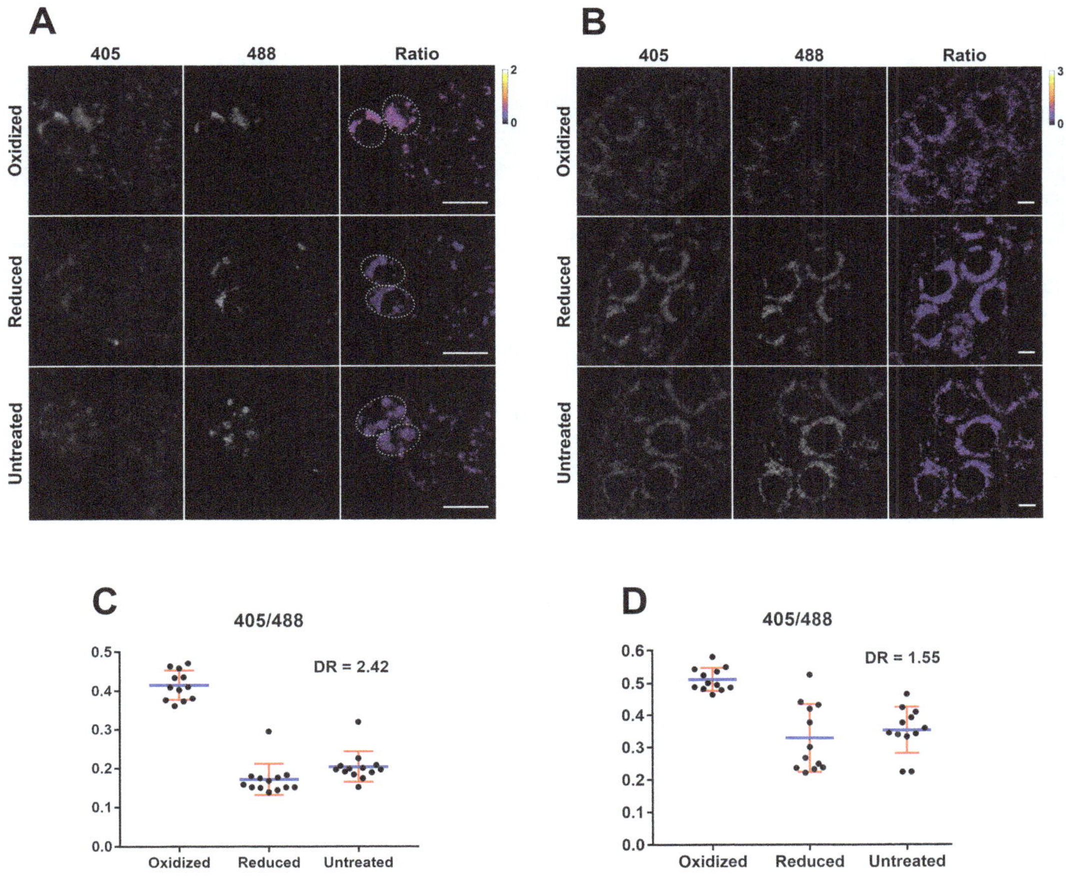

Fig. 2 Mito-roGFP2-Grx1 can monitor redox states in the germline stem cells (GSCs) and their progeny. GSCs (**a**) and nurse cells (**b**) of stage 7 egg chamber showing roGFP2 fluorescence emission after treatment with applied oxidant (DA) or reductant (DTT). (**c**) Ratio of emissions obtained upon excitation of mito-roGFP2-Grx1 in GSCs at 405 and 480 nm in different physiological states and (**d**) in nurse cells of stage 7 egg chamber. DR stands for dynamic range. Scale bar 10 μm (**a**, **b**), and error bars represent SD in red and mean as blue (**c**, **d**)

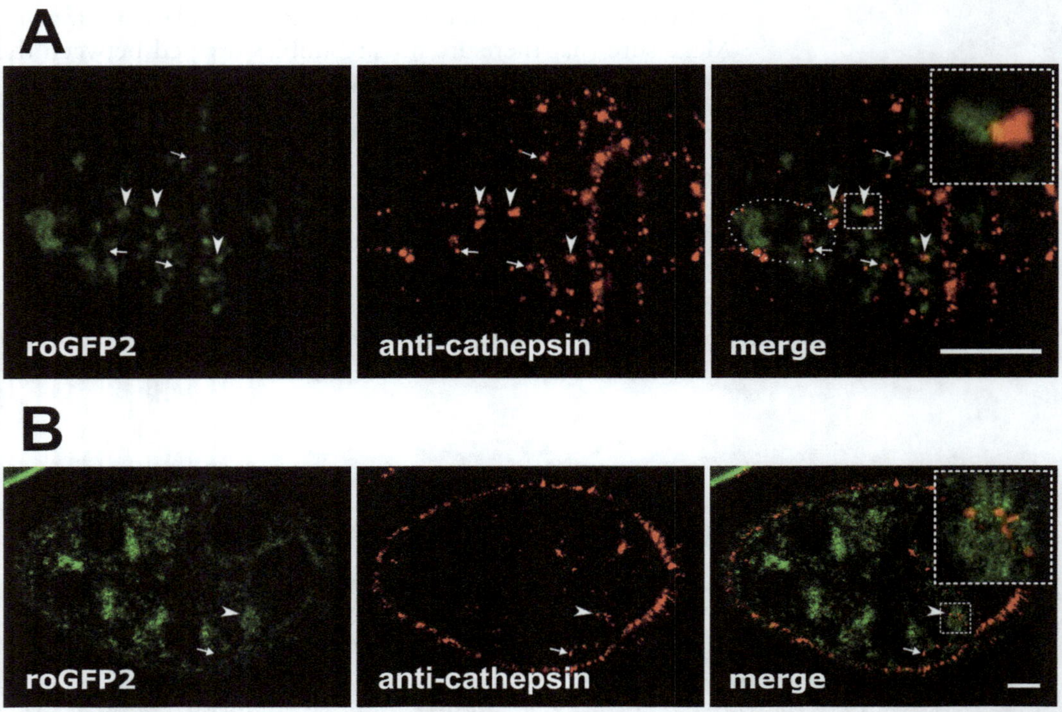

Fig. 3 Mito-roGFP2-Grx1 can be utilized to visualize mitophagy. Mito-roGFP2-Grx1 expression (green) within the germarium (**a**) and a stage 8 egg chamber (**b**) Cathepsin-L marks the lysosomes (red). A few lysosomes are GFP positive (yellow) suggesting the presence of mitochondria within the lysosomal lumen/fusion of autophagosome (containing mitochondrion) (arrow heads). Arrows show GFP negative lysosomes. Inset shows enlarged region of autophagosome (containing mitochondrion)–lysosomal fusion. GSC is marked by dotted oval, scale bar 10 μm

3.4 Image Analysis and Determination of Redox Levels

1. Open the 16 bit image in ImageJ.
2. Process → Subtract Background; rolling ball: 50 pixels.
3. Convert to 32 bit; Image → Type → 32-bit.
4. Manually adjust and set the threshold for the 488 channel; Image → Adjust → Threshold. Set values below the threshold to "not a number" (NaN).
5. Draw an ROI and measure the intensity; Analyze → Measure.
6. In order to keep the same ROI, use ROI manager; Analyze → Tools → ROI manager → select the ROI on the image → it will be shown as an element in the ROI manager → click "Add."
7. Open the 405 channel of the same image → click on the added ROI in the ROI manager → Analyze → Measure.
8. Calculate the 405/488 ratio by dividing the intensity value from 405 channel by the intensity value from the 488 channel of the same image.

9. To generate a ratio image by dividing the 405 channel by the 488 channel, pixel by pixel, use the "Ratio Plus" plugin. For visualization in false colors in Fig. 2, lookup table "Fire" was used.

4 Notes

1. All solutions should be freshly prepared.
2. 1× PBS pH 7.4 is used in preparation of all the solutions. Maintaining pH 7–7.4 is crucial for preventing the quenching of fluorescence.
3. Prepare all the redox reagents on the same day as the treatment.
4. For preparing the three redox reagents, use 1× PBS that has been degassed for at least 1 h. This helps reduce the dissolved oxygen in the solution.
5. Preparing 4% paraformaldehyde fixative freshly ensures better results. 4% PF can also be replaced by 4% formaldehyde solution (devoid of methanol).
6. Although the tissue is fixed post-treatment with redox reagents, plan the imaging the same day as the mounting for best results.

Acknowledgments

We thank Arundhati Bali for technical assistance and members of the Shravage lab for helpful discussions. We thank Dr. K.M. Paknikar, Director, Agharkar Research Institute, Pune, and entire Developmental Biology fraternity for support and access to facilities. This work was supported by grants from DBT-Ramalingaswami Fellowship and DST-SERB grant number ECR/2015/000239. B.V.S. is affiliated to Savitribai Phule Pune University (SPPU), Pune, India and is recognized by SPPU as PhD guide (Biotechnology). KSN is affiliated to Department of Biotechnology, Savitribai Phule Pune University (SPPU).

References

1. Lemasters JJ (2005) Selective mitochondrial autophagy, or mitophagy, as a targeted defense against oxidative stress, mitochondrial dysfunction, and aging. Rejuvenation Res 8:3–5
2. Anding AL, Baehrecke EH (2017) Cleaning house: selective autophagy of organelles. Dev Cell 41:10–22
3. Chen Y, Azad MB, Gibson SB (2009) Superoxide is the major reactive oxygen species regulating autophagy. Cell Death Differ 16:1040–1052
4. Dröge W (2002) Free radicals in the physiological control of cell function. Physiol Rev 82:47–95
5. Dan Dunn J, Alvarez LAJ, Zhang X et al (2015) Reactive oxygen species and mitochondria: a nexus of cellular homeostasis. Redox Biol 6:472–485
6. Shadel GS, Horvath TL (2015) Mitochondrial ROS signaling in organismal homeostasis. Cell 163:560–569

7. Sena LA, Chandel NS (2012) Physiological roles of mitochondrial reactive oxygen species. Mol Cell 48:158–167
8. Gutscher M, Pauleau A-L, Marty L et al (2008) Real-time imaging of the intracellular glutathione redox potential. Nat Methods 5:553–559
9. Zheng M, Åslund F, Storz G (1998) Activation of the OxyR transcription factor by reversible disulfide bond formation. Science 279:1718–1721
10. Østergaard H, Henriksen A, Hansen FG et al (2001) Shedding light on disulfide bond formation: engineering a redox switch in green fluorescent protein. EMBO J 20:5853–5862
11. Albrecht SC, Barata AG, Großhans J et al (2011) In vivo mapping of hydrogen peroxide and oxidized glutathione reveals chemical and regional specificity of redox homeostasis. Cell Metab 14:819–829

Methods in Molecular Biology (2018) 1854: 21–34
DOI 10.1007/7651_2018_166

Published online: 28 June 2018

Induction and Detection of Autophagy in Aged Hematopoietic Stem Cells by Exposing Them to Microvesicles Secreted by HSC-Supportive Mesenchymal Stromal Cells

Rohan S. Kulkarni, Manmohan S. Bajaj, and Vaijayanti P. Kale

Abstract

Autophagy is an important cellular process for maintenance of quality and functionality of cells. This happens through repair and renewal of cellular components like proteins and mitochondria. Reduction in autophagy process in aged hematopoietic stem cells (HSCs) leads to their compromised stemness and self-renewal capacity, and consequently, their applicability in various regenerative therapies also reduces. HSC functions are regulated by their microenvironment, known as "HSC niche," which comprises of mesenchymal stromal cells (MSCs), osteoblasts, endothelial cells, etc. In this niche, the MSCs are known to closely interact with the HSCs, and therefore, they can directly influence the stem cell fate. In our earlier studies, we have demonstrated that young MSCs or aged MSCs rejuvenated by treating them with LY294002, a PI3K inhibitor (rescued aged MSCs), rejuvenate aged HSCs via intercellular transfer of microvesicles (MVs) harboring autophagy-inducing mRNAs.

Here, we describe the protocol for induction of autophagy in aged HSCs by incubating them with microvesicles (MVs) collected from young MSCs or rescued aged MSCs. We also describe the protocols for determination of autophagy levels in these HSCs.

Keywords Autophagy, Hematopoietic stem cells, HSC niche, LC3, Mesenchymal stromal cells, Microvesicles

1 Introduction

In adult mammals, hematopoietic stem cells (HSCs) reside in specialized microenvironment present in the bone cavities, also known as Bone Marrow niche or HSC niche, which governs their functions [1]. By virtue of their close proximity with the HSCs, mesenchymal stromal cells (MSCs) precisely regulate the HSC functionality in these niches [2]. MSCs are known to interact with HSCs in various ways such as cell–cell contact mediated by various adhesion molecules, secretion of soluble factors and extracellular matrix molecules, transfer of reactive oxygen species (ROS), etc. Ratajczak et al., for the first time, demonstrated modulation of HSCs by microvesicles (MVs) secreted by embryonic stem cells

[3]. Furthermore, recently an interesting study showed that HSCs selectively interact with extracellular vesicles (EVs) secreted by HSC-supportive stromal cells, but not with those secreted by non-supportive stromal cells [4]. These reports underscore the importance of EVs secreted by the niche cells in the regulation of HSC functionality.

HSC functionality decreases during aging as reflected in their decreased capacity to reconstitute blood cell chimerism and bone marrow engraftment, along with an increased bias towards myeloid cell formation [5]. Aged HSCs are also known to have reduced autophagy levels. As the subpopulation of aged HSCs having high autophagy flux were found to exhibit higher engraftment potential, as compared to the population having low autophagy flux [6], it became apparent that boosting of autophagy in aged HSCs could boost their engraftment levels. We have demonstrated that incubation of aged HSCs with MVs isolated from HSC-supportive MSCs rejuvenates them and leads to a significant increase in their engraftment capacity by inducing autophagy in them [7].

Here, we describe the protocol for isolation of MVs from MSCs and treatment of aged HSCs with them (Subheading 3.1). Further, we describe the protocol used for induction of autophagy in aged HSCs by exposing them to MVs secreted by young MSCs or with aged MSCs treated with LY294002, a PI3K inhibitor, (rescued MVs) for a brief time period (Subheading 3.2). We also describe the protocol for detection of autophagy levels in aged HSCs after their exposure to young and rescued MVs by immunoblotting (Subheading 3.3) and immunofluorescence (Subheading 3.4) methods.

2 Materials

2.1 Isolation of Microvesicles (MVs) from Mesenchymal Stromal Cells

1. MSCs cultured from bone marrow of young (6–8 weeks) or aged mice (>18 months). (Described in Subheading 3.1.)
2. Iscove's modified Dulbecco's medium (IMDM).
3. Mesenchymal stem cell-qualified fetal bovine serum (MSC-FBS).
4. Phosphate-buffered saline (PBS) 10× solution: NaCl—80.0 g, KCl—2.0 g, Na_2HPO_4—11.5 g, KH_2PO_4—2.0 g, adjust pH to 7.4 by 1 N HCl and make up the volume to 1000 ml with deionized water. Prepare 1× PBS by diluting the 10× stock (1:10) with deionized water just before use.
5. Trypsin phosphate versene glucose (TPVG) solution: 0.1% trypsin, 0.02% EDTA, 0.05% glucose, and phenol red in 1× Dulbecco's PBS.

6. 0.5% (w/v) Bovine serum albumin (BSA; cell culture grade) prepared in IMDM.
7. CO_2 incubator (5% CO_2, humidified).
8. 15 ml Polypropylene centrifuge tubes.
9. Swing bucket centrifuge.
10. 1.5 ml Micro-centrifuge tubes.
11. Temperature controlled benchtop centrifuge (should be able to give a speed of 10,000 × *g*).
12. LY294002 (PI3K inhibitor) 10 mM in DMSO: 10 μM working concentration.
13. Protein assay kit.

2.2 Sorting of Hematopoietic Stem Cells and Their Co-culture with MVs

1. MVs isolated from MSCs (described in Subheading 3.1).
2. Bone marrow cells obtained from long bones (tibia and femur) of young (6–8 weeks) or aged mice (>18 months) for sort purification of LSK (Lin^- $Sca\text{-}1^+$ $c\text{-}Kit^+$) HSCs.
3. Fluorescence-assisted cell sorting (FACS) instrument: any sorter capable of doing sterile sorting.
4. IMDM.
5. Fetal bovine serum (FBS).
6. FACS buffer: 1× PBS supplemented with 2% FBS.
7. Murine recombinant growth factors: stock of 100 μg/ml prepared in 0.1% (w/v) BSA in 1× PBS. Use at the final concentrations of interleukin 3 (IL3): 10 ng/ml, interleukin 6 (IL6): 25 ng/ml, and stem cell factor (SCF): 25 ng/ml.
8. Swing bucket centrifuge.
9. 1.5 ml Micro-centrifuge tubes.
10. 24-Well cell culture grade plates.
11. Bafilomycin-A. Prepare a stock of 20 μM in dimethyl sulfoxide (DMSO): Dilute with IMDM to 200 nM to get working concentration.

2.3 Measuring Autophagy by Western Blot Analysis

1. HSCs cultured with MVs in multi-well plates (typically 24-well plates can be used) (described in Subheading 3.2).
2. 1× PBS (pH 7.4).
3. 1× RIPA (Lysis buffer): Tris–HCl pH 7.4–10 mM, NaCl—100 mM, EDTA—1 mM, EGTA—1 mM, 1% Triton X-100, 0.1% SDS, and 5% sodium deoxycholate. Prepared in deionized distilled water. Store in aliquots up to 6 months at −20 °C.

   Just before use, add the following components to 1 ml of 1× RIPA: 10 μl/ml of 200 mM PMSF stock in ethanol, 10 μl/ml of 10× protease inhibitor cocktail, 5 μl/ml phosphatase inhibitor cocktail I, 5 μl/ml phosphatase inhibitor cocktail II, 10 μl/ml NaF (100 mM), and 10 μl/ml sodium orthovanadate (100 mM Na_3VO_4).

4. Tabletop cooling fixed head centrifuge.
5. Micro-centrifuge.
6. Antibody dilution buffer: 1× Tris-buffered saline (TBS), 0.1% (v/v) Tween 20, and 1% (w/v) BSA. Use freshly prepared.
7. Blocking buffer: TBS containing 0.1% (v/v) Tween 20, and 5% (w/v) BSA. Store at −20 °C and thaw just before use.
8. Running buffer: 25 mM Tris–Cl, pH 8, 192 mM glycine, and 0.1% (w/v) SDS.
9. Stacking buffer: 1.5 M Tris–HCl (pH 6.8).
10. Separating buffer: 1.5 M Tris–HCl (pH 8.8).
11. 6× Sample loading buffer: 10% (w/v) SDS, 20% (v/v) glycerol, 10% (v/v) 2-mercaptoethanol, 1 mM EDTA, 0.002% (w/v) bromophenol blue, 125 mM Tris–Cl, and pH 6.8. Store at −20 °C in working aliquots.
12. TBS: 20 mM Tris–Cl, pH 7.4, and 500 mM NaCl.
13. TBS with Tween 20 (TBST): 20 mM Tris–Cl, pH 7.4, 500 mM NaCl, and 0.1% (v/v) Tween 20. Store at room temperature.
14. Transfer buffer: 25 mM Tris–Cl, pH 8, 192 mM glycine, and 15% (v/v) methanol. Prepare fresh.
15. Pre-stained molecular weight standards and Biotin-labelled molecular weight ladder.
16. Polyvinylidene difluoride (PVDF) membrane.
17. Primary antibodies: rabbit anti-LC3 IgG, and mouse anti-β-actin monoclonal antibody.
18. Secondary antibody: goat anti-rabbit IgG conjugated to horseradish peroxidase (HRP) for rabbit primary antibodies, goat anti-mouse IgG conjugated to HRP for mouse primary antibody, and anti-biotin IgG conjugated to HRP antibody for detection of biotin-labelled molecular weight ladder.
19. Antibody stripping buffer: 2% (w/v) SDS, 0.7% (v/v) 2-mercaptoethanol, 12.5 ml 0.5 M Tris–Cl, and pH 6.8, make volume up to 100 ml by Milli-Q water. Store at RT.
20. HRP substrate.
21. X-ray films.
22. Additional reagents and equipment required for SDS-PAGE and immunoblotting.

2.4 Measuring Autophagy by LC3 Puncta Formation by Immunofluorescence

1. 0.1% w/v Poly-L-lysine solution in water (sterile).
2. HSCs cultured with MVs (described in Subheading 3.2).
3. 2% (w/v) Paraformaldehyde: Dissolve 2 g of paraformaldehyde powder in lukewarm 1× PBS (pH to 7.4). Prepare fresh.
4. 0.1% (v/v) Triton X-100 in 1× PBS.
5. Blocking buffer: 1% (w/v) BSA in 1× PBS.
6. Primary antibody: rabbit anti-LC3 IgG antibody.
7. Secondary antibody: Cy5-conjugated goat anti-rabbit IgG antibody.
8. Aqueous mounting medium: Dissolve 2 mg/ml p-phenylene diamine in PBS containing 50% (v/v) glycerol.
9. 4,6-Diamidino-2-phenylindole (DAPI).
10. Clear nail polish.
11. Glass coverslips, 1.5 mm thickness and 19 mm diameter.
12. Microscopy slides, 76 × 26 mm, lysine coated.
13. Cytospin assembly and instrument.

3 Methods

3.1 Isolation of Micro Vesicles (MVs) from Mesenchymal Stromal Cells

1. Isolate bone marrow cells of 6–8 weeks (young) or ≥18 months (aged) old C57BL/6J (CD45.2) mice by flushing the bone marrow from femur and tibia in 10 ml of IMDM supplemented with 20% MSC–FBS using 26 G needle and 1 ml syringe and seed the cells in 100 mm culture dishes and incubate them at 37 °C in a humidified CO_2 incubator.
2. Remove the non-adherent floating cells after 48 h of incubation and feed the adherent cells with fresh medium every 72 h. Allow the adherent stromal cells to grow till near-confluence before subculturing.
3. At 80–90% confluence, remove the FBS-containing medium and wash the plates with 2 ml TPVG. Add 1 ml of TPVG to the plate and incubate it for a few minutes (1–2 min). Remove TPVG and add IMDM supplemented with 20% MSC–FBS. Flush the cells and collect the cell suspension in 15 ml centrifuge tube. Wash the cells with IMDM supplemented with 20% FBS and seed them in two 100 mm culture plates (split ratio 1:2) in 10 ml of IMDM supplemented with 20% MSC–FBS.
4. Repeat the procedure when cells attain 80–90% confluence (approximately 2–4 days). These cells are referred to as MSCs. Use the MSCs in passage-3 for further experiments.
5. Wash the 70–80% confluent MSCs with plain IMDM and incubate them with IMDM supplemented with 0.5% BSA.

6. For LY294002 (PI3K inhibitor) treatment, incubate aged MSCs with IMDM supplemented with 20% MSC–FBS in the presence of LY294002 (10 μM) for 48 h at 37 °C (*see* **Note 1**). Wash the MSCs with plain IMDM and incubate them with IMDM supplemented with 0.5% BSA in the presence of LY294002 (PI3K inhibitor) (10 μM) for 48 h at 37 °C.
7. After 48 h, collect the conditioned medium in 15 ml sterile polypropylene centrifuge tubes and spin them at 300 × *g* for 10 min at 4 °C to remove floating cells.
8. Collect the supernatant in fresh sterile polypropylene tubes and centrifuge them at 2600 × *g* for 20 min at RT to remove the cell debris and apoptotic bodies (ABs).
9. Collect the supernatant in fresh sterile polypropylene tubes and use it to isolate microvesicles (MVs).
10. Transfer the supernatant collected after removal of apoptotic bodies (**steps 8** and **9**) to 1.5 ml centrifugation tubes and centrifuge at 10,000 × *g* for 1 h at 4 °C in a benchtop centrifuge.
11. Remove the supernatant and resuspend the pellet containing MVs in 2 ml of 1× PBS, transfer to fresh sterile 1.5 ml centrifuge tubes, and spin again at 10,000 × *g* for 1 h at 4 °C in a benchtop centrifuge.
12. Collect the pellet of MVs in 100 μl of 1× PBS or plain IMDM for further experiments. It is imperative to maintain sterility of MVs throughout the procedure (*see* **Note 2**).
13. Estimate the protein content of MVs by BCA method (Protein Assay Kit) (*see* **Note 3**).

3.2 Sorting of Hematopoietic Stem Cells and their Co-culture with Microvesicles

1. Isolate bone marrow mononuclear cells (BM MNCs) from the bone marrow of ≥18 months (aged) old C57BL/6J (CD45.2) mice by flushing the bone marrow in 10 ml of IMDM supplemented with 10% FBS (*see* **Note 4**).
2. Deplete the lineage committed cells by immuno-magnetic separation using biotinylated mouse lineage antibody cocktail and streptavidin-conjugated magnetic beads. Incubate the isolated lineage negative cells with PE-Cy7-conjugated anti-mouse CD117(c-Kit), and PE-conjugated anti-mouse Sca-1 antibodies for 45 min on ice in 1× PBS supplemented with 2% FBS (FACS buffer) (*see* **Note 5**). Wash out the excess antibodies by washing with FACS buffer and resuspend the cells with FACS buffer. Sort the LSK ($Lineage^{-}$ $Sca\text{-}1^{+}$ $c\text{-}Kit^{+}$) cells on ARIA-II FACS machine and collect them in 10 ml of IMDM supplemented with 10% FBS.

3. Wash the sorted cells by resuspending them in IMDM supplemented with 10% FBS and centrifuging them at 300 × *g* for 10 min.
4. Resuspend the cells at a concentration of 100,000 HSCs per 200 μl of IMDM supplemented with 10% FBS and murine growth factors viz. IL-3 (20 ng/ml), IL-6 (50 ng/ml), and SCF (50 ng/ml) (*see* **Note 6**).
5. In another tube, resuspend the MVs isolated from young MSCs (young MVs) or aged LY-treated MSCs (rescued MVs) at a concentration of 10 μg protein equivalent of MVs per 200 μl of IMDM supplemented with 10% FBS.
6. Seed 100,000 (200 μl) HSCs/well of a 24-well tissue culture plate and add 200 μl of MVs (10 μg protein equivalent) to each well. Incubate the cells at 37 °C for 36 h in 5% CO_2 incubator (*see* **Note 7**).

3.3 Detection of Autophagy in Aged Hematopoietic Stem Cells by Western Blot Analyses

This protocol describes an immunoblot assay to detect autophagy in HSCs through conversion of cytosolic LC3-I to lipidated membrane-bound LC3-II [8]. Denatured proteins are resolved by electrophoresis and electrically transferred onto a PVDF membrane. Autophagosome formation is determined by the increase in the amount of LC3-II, relative to a protein loading control (e.g., β-actin).

1. Collect the aged HSCs cultured with young or rescued MVs for 36 h in 1.5 ml centrifuge tubes. Rinse the wells thoroughly with 500 μl of cold 1× PBS and collect the cells in the same centrifuge tubes. Centrifuge the tubes at 1500 × *g* for 5 min to pellet down the HSCs.
2. Wash the cells twice by resuspending the pellet in 1 ml of cold 1× PBS and centrifuging the tubes at 1500 × *g* for 5 min. Remove the supernatant carefully without disturbing the cell pellet.
3. Add 1× RIPA lysis buffer (30 μl) supplemented with protease and phosphatase inhibitor cocktails to the HSC pellet in 1.5 ml centrifuge tube (*see* **Note 8**).
4. Vortex the tube for 30 s, mix well, and incubate the tubes on ice for 20 min to ensure complete lysis.
5. Centrifuge the lysate at 16,000 × *g* for 40 min at 4 °C and collect the supernatant gently into a fresh tube (*see* **Note 9**).
6. Determine the protein concentration by protein estimation kit.
7. Prepare equal protein aliquots (15 μg) and make up the volume to 25 μl with 1× RIPA buffer and add 5 μl of 6× sample buffer (*see* **Note 10**).
8. Heat the samples at 95 °C for 2 min (*see* **Note 11**).

9. Load the samples on 13% SDS PAGE gel and run the electrophoresis at 80 V until the 10 kDa pre-stained molecular marker reaches the bottom of the gel. Take care not to run out the sample (*see* **Note 12**). Also, load biotin-labelled molecular weight ladder along with the samples for determination of molecular weight of proteins of interest after probing with specific antibodies.
10. Transfer the separated proteins onto a PVDF (*see* **Note 13**) membrane using electrical transfer apparatus set at 30 V for overnight at 4 °C (*see* **Note 14**).
11. Next day, dismantle the assembly and remove the membrane. Ensure the transfer of proteins from the gel onto the PVDF membrane by visualizing the transfer of pre-stained protein ladder.
12. Wash the membrane with TBST for 10 min at room temperature.
13. Block the PVDF membrane with 5% BSA in TBST for 3 h, followed by washing with TBST, three times for 10 min each.
14. Dilute rabbit anti-LC3 IgG monoclonal antibody at the recommended concentration (*see* **Note 15**) in 5 ml of antibody dilution buffer and incubate the PVDF membrane with the diluted antibody for 3 h at room temperature on a tabletop rocker (*see* **Note 16**).
15. Wash the membrane three times with TBST for 10 min each at room temperature to remove the unbound/loosely bound/nonspecifically bound primary antibody.
16. Dilute the goat anti-rabbit IgG conjugated to HRP secondary antibody and anti-biotin IgG conjugated to HRP, both diluted appropriately in antibody dilution buffer and probe the PVDF membrane with 5 ml of this antibody for 45 min at room temperature on tabletop rocker.
17. Wash the membrane three times with TBST for 10 min at room temperature to remove the unbound/loosely bound/nonspecifically bound secondary antibody (*see* **Note 17**).
18. Rinse the membrane with Milli-Q water to remove the traces of TBST.
19. Incubate the membrane with 4 ml of chemiluminescent substrate by slowly covering the membrane with substrate for 1 min at room temperature.
20. Drain the excess substrate, keep the membrane in transparent plastic sheets or wrap in saran wrap, and expose to X-ray film in dark for various time spans. Avoid formation of bubbles. Immediately develop and fix the exposed X-ray films using developing and fixing reagents, wash with tap water, air dry and save for analysis (*see* **Note 18**).

21. Wash the membrane once with TBST for 10 min and then incubate in 5 ml of Antibody Stripping Buffer for 30 min at 60 °C to remove the primary and secondary antibodies bound to the membrane (*see* **Note 19**).
22. Wash the membrane three times with TBST for 10 min each, block with 5% BSA in TBST for 3 h, and again wash with TBST for 10 min.
23. The blocked membrane can now be used to detect the levels of β-actin using primary mouse anti-β-actin monoclonal antibody, followed by secondary goat anti-mouse IgG conjugated to HRP. The biotinylated marker can be detected by using anti-biotin IgG conjugated to HRP (*see* **Note 20**) (*see* Fig. 1).
24. Perform densitometric analysis to determine levels of LC3-I, LC3-II, and β-actin in the samples using ImageJ software (NIH). Normalize the values with the input (β-actin).
25. The levels of autophagy can be determined by calculating the ratio of LC3-II:β-actin proteins, and level of autophagic flux in the cells can be determined by ratio of LC3-II:LC-I proteins (for more details refer to Fig. 3E from ref. 7).

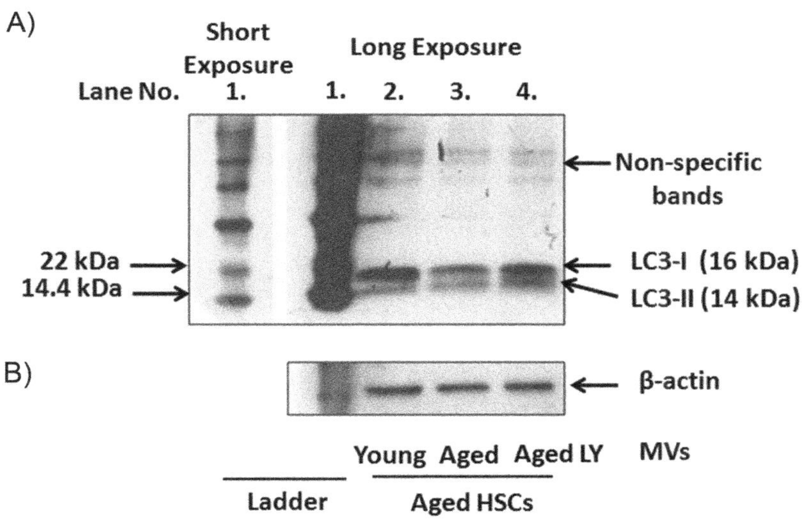

Fig. 1 Detection of autophagy by western blot. (**a**) The image shows western blot analysis of hematopoietic stem cells (HSCs) for detection of LC3. First lane (low exposure/high exposure) shows the biotinylated protein marker (ladder) bands used to identify the exact molecular weight of the protein of interest. Lanes 2–4 show the protein samples from variously treated sets of HSCs as indicated. Two forms of LC3 proteins viz. LC3-I and LC3-II of molecular weight approximately 16 and 14 kDa, respectively, can be clearly seen. (**b**) The image shows the same blot re-probed with anti-β-actin antibody. The same blot used for LC3 detection was re-probed with anti-β-actin antibody after stripping off the anti-LC3 and secondary antibodies. The image shows β-actin protein bands at approximately 42 kDa in each lane

3.4 Detection of Autophagy in Aged HSCs by Immunofluorescence

This protocol describes the detection of MV-mediated autophagy induction in HSCs based on LC3 puncta formation. The level of autophagy directly corresponds to the number and mean fluorescence intensity (MFI) of LC3 puncta per cell.

1. Incubate 100,000 sort-purified aged LSK HSCs with 10 μg protein equivalents of MVs for 36 h in the presence of bafilomycin-A (200 nM), an inhibitor of LC3-II degradation (*see* **Note 21**) or equivalent amounts of DMSO (vehicle control).
2. Collect the co-cultured HSCs as described earlier and wash twice with 1× PBS.
3. Resuspend the HSC pellet in 1× PBS, count and make a suspension of 100,000 cells per 50 μl of volume.

3.5 Prepare the Smears on Slides as Described Below

1. Clean the glass slides and coat them with poly-L-lysine as follows: Take 50 μl of 0.1% poly-L-lysine solution on the slide and spread it uniformly using another slide so as to cover the complete surface. Air dry the slides (*see* **Note 22**).
2. Place the L-lysine coated slides on the cytospin adaptors with coated side facing up. Keep the Thermo-Electron filter cards on them and match the edges of slide to the filter cards. Place the plastic funnel on this assembly and match the holes on filter cards and funnel and clip the clamps to tighten the assembly (*see* **Note 23**).
3. Hold the slide assembly vertical and add 50 μl of cell suspension (approximately 25,000 cells) in each plastic funnel.
4. Spin the assembly in a cytospin instrument at 500 rpm for 2 min.
5. Disassemble the assembly gently, place the slides horizontally, smear facing up, and let the smear air dry.
6. Fix the cells by gently adding 200 μl of 2% PFA on the smear area and incubate the cells at 4 °C for 10 min.
7. Remove the PFA and gently wash the slides with 1× PBS (three times).
8. Permeabilize the cells using 0.2% Triton X-100 in PBS for 10 min.
9. Wash the smear gently three times with 1× PBS (*see* **Note 24**).

3.6 Staining the Cells with LC3 Antibody for Immunofluorescence (IF)

1. Block the smear with 5% (w/v) BSA prepared in 1× PBS. Incubate the slide for 1 h at 4 °C. Remove the blocking solution and wash the smear three times with 1× PBS.
2. Prepare appropriate dilution of anti-LC3 rabbit antibody in 1× PBS supplemented with 1% BSA (*see* **Notes 25** and **26**).

3. Gently add the diluted antibody on the cell smear so as to completely cover the smear and incubate the slide at 4 °C for 1 h in a humidified atmosphere. Do not allow the smear to dry.
4. Wash the smear three times with 1× PBS as discussed earlier (10 min at RT).
5. Clean the area near cell smear with tissue paper, without allowing the smear to dry.
6. Prepare the mixture of appropriately diluted Cy5-conjugateed secondary anti-rabbit antibody and DAPI (1 μg/ml) in 1× PBS supplemented with 1% BSA.
7. Add this mixture drop wise on the smear so as to cover the smear completely and incubate at 4 °C for 1 h in dark (*see* **Note 27**).
8. Wash the smear with 1× PBS three times for 10 min each.
9. Clean the area around the smear using a tissue paper. Do not allow the smear to dry.
10. Put a drop of aqueous mounting medium on the smear area and gently put a coverslip over it. Do not allow formation of air bubbles.
11. Blot the excess mounting medium from the slide using a tissue paper and seal the coverslip with clear nail polish.
12. Keep the slides in a suitable container in dark for acquiring.
13. Acquire the images on a laser scanning microscope at desired resolutions (*see* **Note 28**) (*see* Fig. 2).
14. Score at least ten cells from each field and at least six nonoverlapping fields of each sample set. Calculate percentage LC3 puncta positive cells and determine MFI per cell using ImageJ software (NIH).
15. Data can be represented as number of LC3 puncta/cell and also average LC3 MFI in the sample (for details refer to Fig. 6B from ref. 7).

4 Notes

1. Aged MSCs have increased activation of PI3K/AKT signaling, which affects the composition of their MVs. Inhibition of this signaling rescues the aged MVs by increasing the content of autophagy-inducing mRNAs in them, and thereby positively modulates their effect on aged HSCs [7].
2. The size of MVs is about 200–500 nm. Size and integrity of the isolated MVs can be confirmed by various techniques such as differential light scattering (DLS), scanning electron microscopy (SEM), flow cytometry, etc.

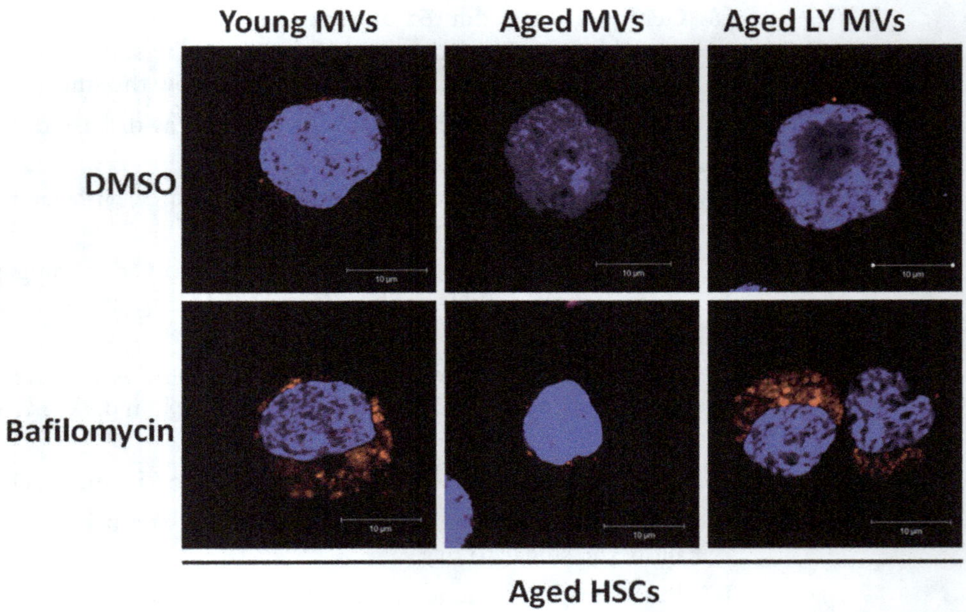

Fig. 2 Detection of autophagy by immunofluorescence (IF). The panel shows images of aged HSCs treated with young, aged, or aged LY MVs with (lower row) or without bafilomycin treatment (Control, upper row). HSCs were stained with Cy5-conjugated anti-LC3antibody (red) to detect LC3 puncta, which appear as dots in the cytoplasmic region. Nuclei are demarked by DAPI (blue). The bar represents 10 μm. The number and intensity of LC3 puncta are directly proportional to the levels of autophagy induction in the cells

3. Protein equivalent yield of MVs is calculated to treat the aged HSCs with equal amount of MVs using BSA as the standard. Aged HSCs are treated with equal amount (i.e., 10 μg equivalents) of young/aged MVs for this experiment.
4. The aged bones are brittle; hence, need to be flushed gently. Application of excess force can result in cracking of bones making it difficult to flush the bone marrow cells.
5. Addition of FBS to PBS is important to reduce the nutritional, physical, and mechanical stress levels experienced by HSCs during the staining procedure.
6. IL3, IL6, and SCF are early acting HSC regulating growth factors that are required for supporting HSC survival during incubation period, and therefore, should be added in the culture medium.
7. Do not disturb the culture plates very often during incubation period so as to avoid dislodging of MVs from the HSCs and to facilitate optimum interaction between HSCs and MVs.
8. Lyse approximately 100,000 HSCs in 10 μl of 1× RIPA buffer. Since HSCs yield very little amount of proteins, the volume of lysis buffer should be kept to minimum so as to obtain desired high protein concentration in the lysate.

9. Centrifugation at 16,000 × *g* ensures the pelleting of unlysed cells, debris, and subcellular components such as nuclei. Supernatant contains the protein fraction, which should be collected very gently to avoid the risk of contamination from contents of the loosely bound pellet.
10. Loading aliquots of equal protein amounts of samples is necessary for precise comparison and calculation of the initial amount of protein loaded and the final amount detected.
11. Also, load 5 μl of biotinylated protein marker mix in one well for exact identification of bands of interest depending on their molecular weight using the known molecular weight bands in this lane.
12. The molecular weights of LC3-I and LC3-II proteins are ~16 and ~14 kDa, respectively. For optimum resolution of the bands, 12–14% polyacrylamide gel should be used. Moreover, samples should not run out of the gel since both, LC3-I and LC3-II run close to the edge of the leading dye, and thus, overrunning of the gel may lead to their loss.
13. Activate the PVDF membrane by treating it with methanol and then wet it in transfer buffer before setting up the transfer.
14. The transfer apparatus generates heat which may affect the binding efficiency of the proteins with the membrane. Thus, it is recommended to perform the transfer in a cold room. Alternatively, the transfer apparatus can be placed in a refrigerator.
15. The binding ability of each antibody to the antigens is different; hence, optimum dilutions should be determined with positive control samples in preliminary experiments.
16. Make sure that the buffer containing antibody is evenly spread over the membrane and covers the entire surface of the membrane to obtain sharp and clear bands.
17. Membrane should be thoroughly washed to completely remove the nonspecifically bound/unbound secondary antibody or loosely bound primary–secondary antibody complex, to ensure good blots with reduced background signals.
18. Alternatively, modern digitalized chemiluminescence image capturing machines can be used to capture digital images of the blot for analysis.
19. Do not incubate the PVDF membrane in antibody stripping buffer for a long time. Excessive incubation in antibody stripping buffer may result in detaching of proteins from PVDF membrane, hampering the detection of subsequent proteins.
20. Follow the same procedure for detection of β-actin as described above for LC3 proteins using western blotting (chemiluminescence detection).

21. Bafilomycin A inhibits fusion of autophagosome and lysosome complex and thereby inhibits LC3-II degradation, thus, enabling its detection. Levels of LC3-II are directly correlated with the autophagy levels.
22. Poly-L-lysine coating is required to decrease the detachment of HSCs from the slides during further procedure.
23. Be careful to precisely match the plastic funnel opening to the hole in the filter cards to ensure proper smear formation.
24. Slides having fixed smears can be stored in 1× PBS at 4 °C but should be processed as soon as possible.
25. Make sure that the primary antibody used for the detection of LC3 puncta is IF qualified.
26. Include the proper isotype control for primary and secondary antibodies used as controls for nonspecific binding of antibodies.
27. Fluorochrome tagged to the secondary antibody is photosensitive and quenches when exposed to bright light. Hence, incubation of cells with secondary antibodies should be done in dark using a light impermeable box. Further procedure should also be carried out in deemed light.
28. LC3 puncta are visible in cytoplasmic region of the HSCs. Higher number of puncta/cell or higher MFI of the LC3 puncta indicates higher autophagy levels in these cells. DAPI is used to demarcate nuclei. It also helps for focusing the cells of interest (Excitation/Emission (nm): 358/461).

References

1. Lymperi S, Ferraro F, Scadden DT (2010) The HSC niche concept has turned 31. Ann N Y Acad Sci 1192(1):12–18
2. Méndez-Ferrer S, Michurina TV, Ferraro F et al (2010) Mesenchymal and haematopoietic stem cells form a unique bone marrow niche. Nature 466(7308):829
3. Ratajczak J, Miekus K, Kucia M et al (2006) Embryonic stem cell-derived microvesicles reprogram hematopoietic progenitors: evidence for horizontal transfer of mRNA and protein delivery. Leukemia 20(5):847
4. Stik G, Crequit S, Petit L et al (2017) Extracellular vesicles of stromal origin target and support hematopoietic stem and progenitor cells. J Cell Biol. https://doi.org/10.1083/jcb.201601109
5. Geiger H, Rudolph KL (2009) Aging in the lympho-hematopoietic stem cell compartment. Trends Immunol 30(7):360–365
6. Ho TT, Warr MR, Adelman ER et al (2017) Autophagy maintains the metabolism and function of young and old stem cells. Nature 543 (7644):205
7. Kulkarni R, Bajaj M, Ghode S et al (2018) Intercellular transfer of microvesicles from young mesenchymal stromal cells rejuvenates aged murine hematopoietic stem cells. Stem Cells 36 (3):420–433
8. Klionsky D, Abeliovich H, Agostinis P et al (2008) Guidelines for the use and interpretation of assays for monitoring autophagy in higher eukaryotes. Autophagy 4:151–175

Methods in Molecular Biology (2018) 1854: 35–43
DOI 10.1007/7651_2017_64

Published online: 27 August 2017

Skeletal Muscle Lysosomal Function via Cathepsin Activity Measurement

Kristyn Gumpper, Matthew Sermersheim, Michael X. Zhu, and Pei-Hui Lin

Abstract

Muscle wasting or cachexia is commonly associated with aging and many diseases such as cancer, infection, autoimmune disorders, and trauma. Decrease in muscle mass, or muscle atrophy, is often caused by dysfunction of protein proteolytic systems, such as lysosomes, which regulate protein turnover and homeostasis. Lysosomes contain many hydrolases and proteases and, thus, represent the major organelle that control protein turnover. Recently, lysosomes have emerged as a signaling hub to integrate cellular functions of nutrient sensing and metabolism, autophagy, phagocytosis, and endocytosis, which are all related to tissue homeostasis. In this chapter, we describe the protocol used to measure lysosomal proteinase (cathepsins) activity in the skeletal muscle. A better understanding of lysosomal function in muscle homeostasis is critical in developing new therapeutic approaches to prevent muscle wasting.

Keywords Atrophy, Autophagy, Enzyme kinetics, Fluorimeter, Muscle acid lysates (MAL), Protein degradation, Skeletal muscle function

1 Introduction

Skeletal muscle is integral to human health for both movement and metabolism, and is maintained by a balance of protein synthesis and degradation [1, 2]. Upon disuse, such as during bed rest, aging, or due to diseases like diabetes mellitus or cancer, disruption of protein homeostasis and subsequent atrophy of tissue cause skeletal muscle to lose between 6 and 24% of its mass and strength [3, 4]. Atrophy is the removal of proteins via degradation without subsequent protein synthesis. In skeletal muscle, atrophy is often characterized as a loss of muscle mass caused by the increase in one or more of the four canonical protein degradation pathways: calcium-dependent cysteine proteases (calpains), cysteine-aspartic acid proteases (caspases), lysosomal cysteine proteases (cathepsins), and ubiquitin mediated proteasomal degradation [5]. For the purpose of this protocol, we will focus on the lysosomal cysteine protease (cathepsins) protein degradation pathway for studying skeletal muscle atrophy.

In its conventional role, lysosomes are viewed as digestive organelles for macromolecule turnover. Lysosomes are membrane bound acidic organelles that contain a plethora of enzymes which breakdown molecules that are no longer needed by the cell via hydrolysis. Recently, lysosomes have emerged as a signaling center that integrate many cellular processes, such as nutrient sensing, metabolism, and autophagy, which are essential to cell and tissue homeostasis [6, 7]. Moreover, lysosomal membranes contain many channels and transporters that maintain luminal ion homeostasis and are essential to regulate digestive enzyme activity, vesicular trafficking, and response to cellular metabolic stress. Impaired lysosomal function is often associated with cancer, neurodegenerative disorders, and lysosomal storage diseases [6–8].

Autophagy is a nonselective process by which cytoplasmic contents, such as damaged organelles and aged protein aggregates, are degraded. Targeted cytoplasmic content is engulfed by double membrane bound vesicles, known as autophagosomes, and are subsequently delivered to lysosomes for bulk degradation within the cell. Lysosomes fuse with autophagosomes to form autolysosomes, allowing the acidic environment and enzymes within the lysosome to interact specifically with the biomolecules targeted for degradation [9, 10]. Recent studies of autophagy in skeletal muscle of cancer patients with and without cachexia have examined the expression of proteins and their mRNA upstream of lysosomal formation [9]. Other studies have used fluorescent dyes such as LysoTracker Green to monitor the movement of lysosomes in conjunction with other fluorescent dyes during different stages of autophagy. However, although these studies provide detailed information on lysosome biogenesis and trafficking, they do not directly assess lysosome function.

Cysteine cathepsins are a group of 15 proteins, representing the most abundant enzymes present in the lysosome that mediate proteome homeostasis [8, 11]. Cathepsins are ubiquitiously expressed with the exception of cathepsin K (osteoclasts, epithelial cells, and synovial fibroblasts), cathepsin S (antigen presenting cells such as dendritic cells and B-cells), cathepsin V (thymus and testis), and cathepsin W ($CD8^+$ lymphocytes and NK cells) [8, 12]. Since cathepsins regulate protein turnover within the lysosome, they provide an excellent readout for lysosomal function. Cathepsins function by binding target proteins in their active site cleft and hydrolyzing target peptides preferentially after basic or hydrophobic residues [13]. Early studies of lysosomal degradation kinetics were performed via radioactive labeling of peptides or proteins with [^{14}C] acetylated hemoglobin, a substrate for cathepsins D and L [14]. Alternative methods to quantify cathepsin activity also exist, such as using fluorogenic substrates in conjunction with immunofluorescent microscopy [15], measuring the fluorescence extinction coefficient of trichloroacetic acid (TCA) extracts [16], or zymography [17]. To study individual cathepsin activity, subtype specific substrates, such as Ac-RR-AFC (Arg-Arg-7-amino-4-trifluoromethylcoumarin) for cathepsin B, could be applied [18].

In this protocol, we demonstrate extraction of lysosomes from mouse skeletal muscle and the use of a sensitive omnicathepsin fluorogenic substrate assay to measure cathepsin activities. This cathepsin activity assay protocol measures the total activities of cathepsins B, C, F, K, L, O, S, and X and is not specific for a specific cathepsin subtype. The substrate Z-FR-AMC (Z-Phe-Arg-AMC) is a small peptide attached to an AMC group (7-amino-4-methylcoumarin) that fluoresces at 465 nm when cleaved via hydrolysis from the peptide and excited at 360 nm. To ensure that the activity of cathepsins is assayed via AMC fluorescence, a parallel set of samples are treated with the cathepsin inhibitor, E-64-D (L-trans-Epoxysuccinyl-leucylamido(4-guanidino)butane), as a negative control. It is important to note that all cathepsins can be inhibited by E-64-D, with the exception of cathepsin C which is only partially inhibited. Rates of cathepsin activity are calculated based on the amount of Z-FR-AMC cleavage, derived from a standard curve of free AMC. Additionally, cathepsin specific activity can be measured as *V*max (RFU/time) kinetics. Compared to the inherent risk of working with radioactive chemicals and the semi-quantitative nature of fluorescent microscopy and zymography, this protocol accurately and consistently measures cathepsin activity and is readily adaptable for specific cathepsin activity [19].

2 Materials

1. Tissue homogenizer with plastic Omni tips (OMNI International).
2. Centrifuge (either temperature controlled or placed in a refrigerator at 4 °C).
3. Sonicator (Branson Digital Sonifier).
4. 96-Well plate, opaque-walled, flat-bottom plates (available from several suppliers).
5. Fluorimeter or microplate reader capable of reading fluorescence and temperature control (e.g., FlexStation 3 from Molecular Device): Pre-warmed to 37 °C.
6. Cytosol lysis buffer: 20 mM Tris–HCl, 1 mM EDTA, 1 mM EGTA, 1% glycerol, and 2 mM dithiothreitol (DTT, freshly prepared) (pH 7.8) (*see* **Notes 1** and **2**).
7. Acid lysis buffer: 200 mM Na-acetate, 50 mM NaCl, and 0.1% Triton X-100 (pH 5.0) (*see* **Note 3**).
8. Cathepsin assay buffer: 100 mM Na-acetate, 120 mM NaCl (pH 5.0), and 1 mM EDTA (pH 5.5).

9. E-64-D, Enzo Life Sciences (BML-PI107): 10 mM stock in DMSO stored at −20 °C. Working solution made fresh at 325 μM by diluting the 10 mM stock 1:30.8 in the Cathepsin Assay Buffer. The working concentration during preincubation period is 50 μM.
10. Z-FR-AMC (benzyloxycarbonyl-Phe-Arg-(7-amino-4-methylcoumarin)), Enzo Life Sciences (BML-P139): 50 mM substrate stock in DMSO stored at −20 °C in 2.4 μL aliquot in 0.5 mL amber tubes to protect from light (*see* **Note 4**).
11. AMC (7-amino-4-methylcoumarin) fluorescence reference standard (ThermoFisher Scientific, A191, mol. weight 175): 32 mM stock in DMSO stored at −20 °C, to be used for free AMC standard curves (*see* Sect. 3.2).

3 Methods

3.1 Lysate Extraction

Carry out all procedures on ice unless otherwise specified.

1. Weigh muscle and mince in 3 volumes of ice cold Cytosol Lysis Buffer.
2. Homogenize with a fresh Omni tip for each sample.
3. Centrifuge at 13,000 × *g* for 30 min at 4 °C.
4. Remove the supernatant. Retain the pellet for lysosome extraction and cathepsin activity assay. The supernatant is mainly soluble cytoplasmic proteins and other soluble cellular debris, which can be stored in a separate tube at −20 °C for other experiments. The pellet contains membrane-bound organelles, including lysosomes.
5. To break open the pellets containing lysosomes, add two volumes of acid lysis buffer to the pellet and sonicate on ice for 3 min with 10 s to 5 s "on-off" cycles.
6. Centrifuge again at 13,000 × *g* for 30 min at 4 °C.
7. Collect this supernatant (muscle acid lysates, MAL) in a labeled 1.5 mL tube for the cathepsin activity assay.
8. Measure protein concentration of MAL with assays such as BCA, Lowry, or Bradford assays. Either use the samples immediately or store them at −80 °C for up to 1 month (*see* **Note 5**).
9. (Optional) Confirm success of protein extraction in MAL via a 10% SDS-PAGE, with the gel stained by Colloidal Blue (*see* **Note 6** and Fig. 1).

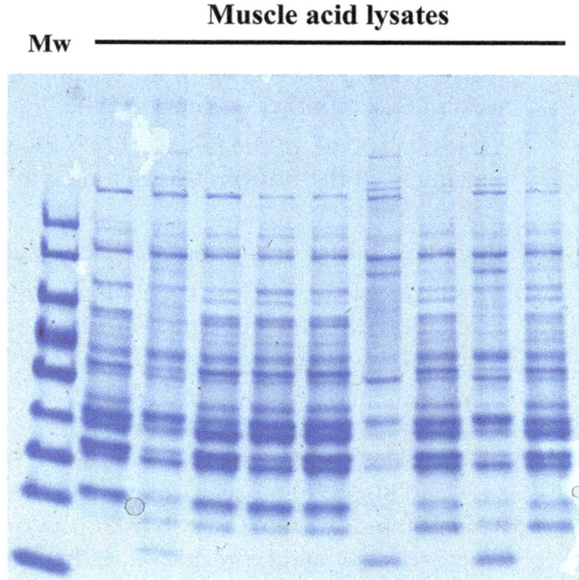

Fig. 1 The acid/lysosome extracts (3 μg/lane) from murine gastrocnemius muscle were separated on a 10% SDS-PAGE and stained with Colloidal Blue. *Mw* molecular weight marker (*lane 1*)

3.2 Standard Curve

Store Z-FR-AMC properly to prevent autolysis which could cause auto-fluorescence and interfere with the assay readout. To take into account of instability of Z-FR-AMC during storage, the standard curve should be derived from a calibration curve made from a 1:1 mixture of serial dilutions of free AMC (final 0–8 μM) and a constant concentration of Z-FR-AMC (final 20 μM).

1. Thaw a vial containing 32 mM free AMC stock solution. Dilute 1:1,000 with the cathepsin assay buffer to make a 32 μM free AMC solution.
2. Make serial dilutions of free AMC to final concentrations of 16, 8, 4, 2, 1, 0.5, 0.25, and 0 μM with a minimum final volume of 120 μL.
3. Thaw a vial containing 2.4 μL 50 mM Z-FR-AMC stock and dilute with the cathepsin assay buffer to make a 3 mL 40 μM Z-FR-AM solution of working substrate.
4. Label 8 calibration tubes for AMC/Z-FR-AMC mixture for 8, 4, 2, 1, 0.5, 0.25, 0.125, and 0 μM free AMC.
5. Mix 120 μL free AMC (16 μM) with 120 μL Z-FR-AMC (40 μM) to make the highest calibration point of 8 μM free AMC with 20 μM Z-FR-AMC.
6. Follow the same dilution scheme to mix the next free AMC (8 μM) with Z-FR-AMC (40 μM) to make the next calibration point of 4 μM free AMC with 20 μM Z-FR-AMC.

7. Follow the same dilution scheme to make all calibration points with the lowest one being 0 μM free AMC with 20 μM Z-FR-AMC. This is the background reading to be subtracted from readouts from other calibration tubes.
8. Transfer 100 μL of the free AMC/Z-FR-AMC mixture into an opaque wall flat bottom 96-well plate in duplicates from each calibration tube.
9. Include a buffer only sample as background reading (blank).
10. Incubate at 37 °C for 1 h and measure fluorescence with excitation 380 nm and emission 460 nm.
11. Generate an AMC standard calibration curve. A typical AMC standard curve is shown in Fig. 2.

3.3 Cathepsin Activity Assay

A pilot experiment is recommended using different amounts (0, 3, 7.5, 15, and 30 μg) of muscle acid lysates (MAL) and 20 μM Z-FR-AMC in 100 μL cathepsin assay buffer in a 96-well format. We have obtained sensitive and reproducible cathepsin assay measurements with 15 μg MAL and thus this is recommended accordingly [19].

1. Pre-warm the fluorimeter to 37 °C for optimal cathepsin activity assay.
2. Dilute 70 μg MAL to 0.6 μg/μL in the cathepsin assay buffer with a minimum final volume of 120 μL per sample.
3. Place 55 μL (equivalent to 33 μg MAL) into two sets of tubes labeled "A"—without E-64-D, and "B"—with E-64-D for each sample.
4. Add 10 μL cathepsin assay buffer to set A tubes and 10 μL 325 μM E-64-D to set B tubes (the final concentration of E-64-D is 50 μM) and incubate at 37 °C for 15 min. This should inactivate all cathepsin activity in set B samples.

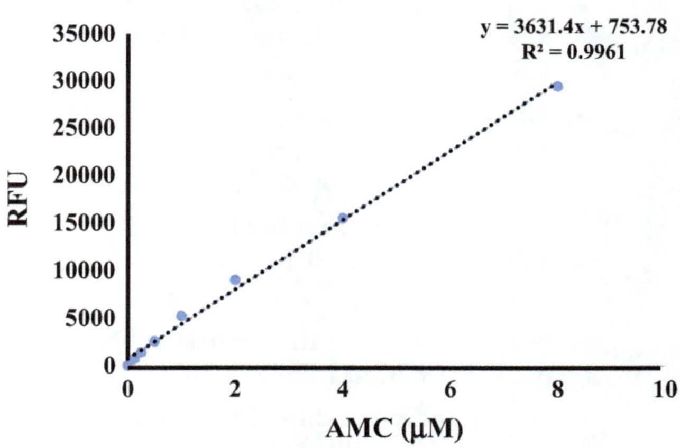

Fig. 2 A typical AMC standard calibration curve with relative fluorescence readout. *RFU* relative fluorescence units

5. Place 30 μL of the above solution (equivalent to 15 μg MAL) into duplicated wells in a 96-well (opaque-walled, flat bottom) plate and add additional 20 μL cathepsin assay buffer into each well.
6. Dilute Z-FR-AMC stock (50 mM) to a 40 μM working solution. To do this, mix 2.4 μL stock with 3 mL of the cathepsin assay buffer (*see* **Note 4**).
7. Add 50 μL of the 40 μM Z-FR-AMC working solution to each well. The final Z-FR-AMC substrate concentration is 20 μM, with 100 μL cathepsin assay buffer per well.
8. Incubate the plate in the fluorimeter at 37 °C. Read at excitation of 380 nm and emission of 460 nm every 3 min for 60 min. Record the intensity values, which will give a curve describing the kinetics of cathepsin activity (*see* Sect. 3.4 and Fig. 3).

3.4 Cathepsin Activity Calculation

The data obtained can be displayed in two ways.

1. Relative fluorescence units. The relative cathepsin activity is calculated based on the amount of free AMC [in relative fluorescence unit (RFU)] released from Z-FR-AMC by enzymatic cleavage, determined by comparing to the standard calibration curve of AMC (Fig. 2). The RFU values of all samples should be corrected by subtracting the value of the blank. The mean RFU value for each sample in duplicates is then calculated.
2. Relative enzyme kinetics (*V*max). The rate of cathepsin enzyme kinetics is calculated as the change in free AMC (RFU) per

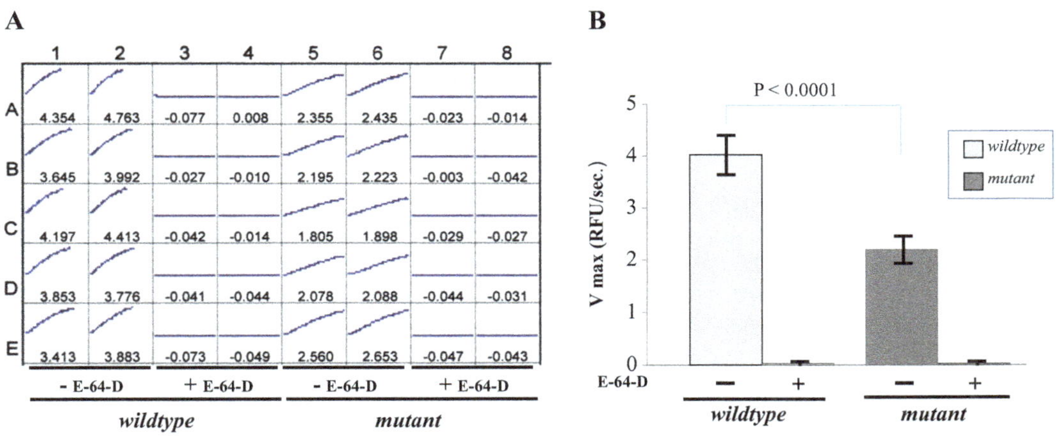

Fig. 3 Acid protein extracts, derived from gastrocnemius muscles of wild type (wt) or mutant mice, were assayed simultaneously to measure cathepsin proteinase kinetics with 20 μM Z-FR-AMC fluorogenic peptide substrate. Specific cathepsin activity was assessed using the cysteine protease inhibitor E-64-D (50 μM). (**a**) A typical readout from a fluorimeter with plots of enzymatic kinetics (*V*max, RFU/s) for the MALs. (**b**) A bar graph representation of the aggregated results of the enzyme kinetics expressed as RFU/s

second resulting from Z-FR-AMC cleavage in MAL for each sample treated with or without E-64-D (Fig. 3). This can also be displayed as RFU/μg total protein/min.

4 Notes

1. The extraction buffer does not contain any detergent to ensure that membrane-bound lysosomes stay intact through the first few steps.
2. Prepare DTT fresh by adding 3.1 mg to 10 mL buffer to a final concentration of 2 mM. DTT alkalizes over time and the reducing power of DTT is limited to pH 7.0 and above.
3. Acid lysis buffer is acidic to match the acidic environment of the lysosome lumen, ideal for most cathepsin activities. The triton X-100 detergent along with sonication will break apart the lysosome membrane.
4. Store Z-FR-AMC aliquots individually to reduce the number of freeze-and-thaw times. A 2.4 μL 50 mM Z-FR-AMC tube makes 3 mL of the 40 μM working substrate solution.
5. A typical yield of acid/lysosome extraction is 0.7–1.6 μg/μL.
6. A typical acid/lysosome extraction on a 10% SDS-PAGE gel is shown in Fig. 1.

Acknowledgments

This work was supported by NIH grant R01GM092759 (to M.X. Z) and an Ohio State University intramural Lockwood Research grant (to P.H.L).

References

1. Bonaldo P, Sandri M (2013) Cellular and molecular mechanisms of muscle atrophy. Dis Model Mech 6:25–39
2. Sakuma K, Aoi W, Yamaguchi A (2017) Molecular mechanism of sarcopenia and cachexia: recent research advances. Pflugers Arch 469 (5-6):573–591
3. Narici MV, de Boer MD (2011) Disuse of the musculo-skeletal system in space and on earth. Eur J Appl Physiol 111:403–420
4. Chopard A, Hillock S, Jasmin BJ (2009) Molecular events and signalling pathways involved in skeletal muscle disuse-induced atrophy and the impact of countermeasures. J Cell Mol Med 13:3032–3050
5. Bialek P, Morris C, Parkington J, St Andre M, Owens J, Yaworsky P, Seeherman H, Jelinsky SA (2011) Distinct protein degradation profiles are induced by different disuse models of skeletal muscle atrophy. Physiol Genomics 43:1075–1086
6. Xu H, Ren D (2015) Lysosomal physiology. Annu Rev Physiol 77:57–80
7. Xiong J, Zhu MX (2016) Regulation of lysosomal ion homeostasis by channels and transporters. Sci China Life Sci 59:777–791
8. Stoka V, Turk V, Turk B (2016) Lysosomal cathepsins and their regulation in aging and neurodegeneration. Ageing Res Rev 32:22–37

9. Aversa Z, Pin F, Lucia S, Penna F, Verzaro R, Fazi M, Colasante G, Tirone A, Rossi Fanelli F, Ramaccini C, Costelli P, Muscaritoli M (2016) Autophagy is induced in the skeletal muscle of cachectic cancer patients. Sci Rep 6:30340
10. Eskelinen EL, Saftig P (2009) Autophagy: a lysosomal degradation pathway with a central role in health and disease. Biochim Biophys Acta 1793:664–673
11. Reiser J, Adair B, Reinheckel T (2010) Specialized roles for cysteine cathepsins in health and disease. J Clin Invest 120:3421–3431
12. Mohamed MM, Sloane BF (2006) Cysteine cathepsins: multifunctional enzymes in cancer. Nat Rev Cancer 6:764–775
13. Choe Y, Leonetti F, Greenbaum DC, Lecaille F, Bogyo M, Bromme D, Ellman JA, Craik CS (2006) Substrate profiling of cysteine proteases using a combinatorial peptide library identifies functionally unique specificities. J Biol Chem 281:12824–12832
14. Schroter J, Schott KJ, Purtill MA, Neuhoff V (1986) Lysosomal protein degradation in experimental hyperphenylalaninaemia. J Inherit Metab Dis 9:273–282
15. Creasy BM, Hartmann CB, White FK, McCoy KL (2007) New assay using fluorogenic substrates and immunofluorescence staining to measure cysteine cathepsin activity in live cell subpopulations. Cytometry A 71:114–123
16. Rossman MD, Maida BT, Douglas SD (1990) Monocyte-derived macrophage and alveolar macrophage fibronectin production and cathepsin D activity. Cell Immunol 126:268–277
17. Wilder CL, Park KY, Keegan PM, Platt MO (2011) Manipulating substrate and pH in zymography protocols selectively distinguishes cathepsins K, L, S, and V activity in cells and tissues. Arch Biochem Biophys 516:52–57
18. Jang BG, Choi BY, Kim JH, Kim MJ, Sohn M, Suh SW (2013) Impairment of autophagic flux promotes glucose reperfusion-induced neuro2A cell death after glucose deprivation. PLoS One 8:e76466
19. Lin PH, Duann P, Komazaki S, Park KH, Li H, Sun M, Sermersheim M, Gumpper K, Parrington J, Galione A, Evans AM, Zhu MX, Ma J (2015) Lysosomal two-pore channel subtype 2 (TPC2) regulates skeletal muscle autophagic signaling. J Biol Chem 290:3377–3389

Methods in Molecular Biology (2018) 1854: 45–53
DOI 10.1007/7651_2017_65

Published online: 17 August 2017

Autophagy in Adipocyte Differentiation

Zhipeng Tao, Longhua Liu, Louise D. Zheng, and Zhiyong Cheng

Abstract

Adipose tissue regulates metabolic homeostasis by acting as an endocrine organ and energy reservoir. Adipose tissue development and functional maintenance are dependent on adipocyte differentiation, in which autophagy plays an important role. It has been shown that autophagy deficiency dampens adipocyte differentiation, compromises adipose tissue development, dysregulates adipocytokine secretion, and even causes sudden death in young animals. Therefore, accurate assessment of autophagy in adipocyte is critical for the study of adipose biology or pathology of metabolic diseases. In this chapter, we described the procedure of autophagy analysis during adipocyte differentiation, and discussed the power of steady-state autophagy protein (e.g., beclin 1, LC3, and p62) levels versus autophagy flux to reflect autophagy activity.

Keywords Adipocyte differentiation, Adipose development and function, Autophagy, Endocrine, Metabolism

1 Introduction

Adipose tissue has been recognized as a critical endocrine organ in mammals [1–3]. In addition to storing energy, adipose tissue can secret adipocytokines that regulate inflammation, cell signaling, and metabolism [1–3]. Positive energy balance increases adiposity, which may lead to hypoxia because of rapid expansion of adipose tissue, thereby dysregulating adipocyte function, adipocytokine secretion, and metabolic homeostasis [4, 5]. Nevertheless, heathy expansion of adipose tissue after adipocyte differentiation is critical for adipose tissue maintenance and function [4, 5]. Recently, we found that adipocyte differentiation and lipid droplet expansion were dependent on autophagy, a common mechanism that cells adapt for cellular remodeling and recycling [6–9]. Suppression of autophagy dampens adipocyte differentiation, reduces fat mass, and induces browning of white adipose tissue [10, 11]. However, embryonic deletion of autophagy genes seems to compromise adipose tissue development and function, resulting in a substantial death rate in model animals at young ages [10]. These findings suggest that autophagy is a central regulator of adipocyte differentiation, adipose tissue development, and functional maintenance.

Autophagy process consists of multiple steps, i.e., initiation, vesicle elongation, vesicle completion, fusion, and degradation [7, 12]. Among the identified autophagy proteins, beclin 1 is a critical component of the initiation complex, while LC3-II is found to participate in vesicle elongation, vesicle completion, and fusion [7, 12]. In addition, LC3-II can be selectively degraded by autolysosome along with adaptor protein p62 such that the turnover of these two proteins reflects autophagy flux [7, 12]. Given the highly dynamic nature and multi-step process of autophagy, it has been recognized that substrate turnover (or autophagy flux) other than steady-state parameters (e.g., steady-state levels of autophagy genes or their encoded proteins, or autophagosome number counted with electron microscopy and fluorescence microscopy) can reveal factual autophagy activity [12, 13]. In this chapter, we described the protocol to determine the steady-state levels of autophagy gene-encoded proteins, as well as autophagy flux during adipocyte differentiation, contrasting their abilities to reflect factual autophagy activity in adipocytes.

2 Materials

2.1 Cell Culture

1. Equipment and supplies: 10-cm tissue culture dishes, 6-well tissue culture plates, 50-ml sterile conical tubes, sterile tips, pipettes, light microscope, laboratory CO_2 water-jacketed incubators benchtop centrifuge, and biosafety cabinet.
2. 3T3-L1 cell line (CL-173) purchased from ATCC (Manassas, VA, USA).
3. Stromal vascular fraction (SVF) cells isolated freshly from subcutaneous white adipose tissue (sWAT) as described previously [14].
4. Phosphate-buffered saline.
5. Basal medium for 3T3-L1 cells: Dulbecco's modified Eagle's (DMEM) medium, 10% fetal bovine serum (FBS), and 1× Pen/Strep.
6. Basal medium for SVF cells: DMEM/F12, 10% fetal bovine serum (FBS), and 1× Pen/Strep.

2.2 Adipocyte Differentiation

1. Equipment and supplies: 6-well tissue culture plates, sterile tips, pipettes, inverted microscope, laboratory CO_2 water-jacketed incubators, and biosafety cabinet.
2. Differentiation media (DM) for 3T3-L1 cells: DMI–DMEM supplemented with 10% FBS, P/S (1×), IBMX (0.5 mM), dexamethasone (1 μM), insulin (1 μg/ml), and rosiglitazone (2 μM). DMII–DMEM supplemented with 10% FBS, P/S (1×), and insulin (1 μg/ml).

3. Differentiation medium for SVF cells: DMEM/F12 media containing 10% FBS, 1× Pen/Strep, dexamethasone (5 μM), insulin (0.5 μg/ml), IBMX (0.5 mM), and rosiglitazone (1 μM).
4. Maintenance medium for 3T3-L1 cells: DMEM media containing 10% FBS, 1× Pen/Strep.
5. Maintenance medium for SVF cells: DMEM/F12 media containing 10% FBS, 1× Pen/Strep, and insulin (0.5 μg/ml).

2.3 Autophagy Flux Assay

1. Bafilomycin A1 (inhibitor of autophagosome acidification, 0.1 mM in DMSO as the stock solution).
2. Leupeptin (inhibitor of lysosomal proteases, 10 mg/ml in DMSO as the stock solution).

2.4 Western Blotting

1. Equipment and supplies: 50-ml sterile conical tubes, sterile tips, 2 ml sterile tubes, pipettes, benchtop centrifuge, microplate reader, heat block (Benchmark).
2. Phosphate buffered saline (PBS).
3. Bullet Blender® (Next Advance, Inc.).
4. DC protein assay kits (Bio-Rad).
5. PLC lysis buffer: 30 mM Hepes, pH 7.5, 150 mM NaCl, 10% glycerol, 1% Triton X-100, 1.5 mM $MgCl_2$, 1 mM EGTA, 10 mM NaPPi, 100 mM NaF, 1 mM Na_3VO_4 supplemented with protease inhibitor cocktail (Roche), and 1 mM PMSF (freshly added right before use).
6. Loading buffer (5×): Tris–HCl (pH 6.8) 0.25 M, glycerol 25%, 5% SDS, bromophenol blue 0.25%, and 0.5 M DTT.
7. Running buffer (1×): 25 mM Tris-Base, 192 mM glycine, and 0.1% SDS (pH 8.3).
8. Transfer buffer: 25 mM Tris-Base, 192 mM glycine (pH 8.3), and 15–20% methanol.
9. Washing buffer: 50 mM Tris–HCl, 150 mM NaCl, and 0.1% Tween 20 (pH 7.4).
10. Antibodies: Beclin 1 (Cat No: MABN16) from EMD Millipore. p62 (Cat No: 5114s) from Cell Signaling Technology; LC3 (cat No: L7543) from Sigma-Aldrich; and GAPDH (Cat No: MA5-15738) from ThermoFisher Scientific.

3 Methods

3.1 Differentiation of 3T3-L1 Cells

1. Seed 3T3-L1 cells in 10-cm dishes and then subculture in 6-well plates with basal media.
2. Change the media every 2 days until the cells reach confluence (day 0).

3. Change the media one more time and maintain the cells in basal media till day 2.
4. At the end of day 2, replace basal media with differentiation medium I (DMI).
5. At the end of day 4, replace DMI with differentiation medium II (DMII).
6. At the end of day 6, replace DMII medium with maintenance medium.
7. Change maintenance medium every 2 days until day 12 when the cells are fully differentiated, which is confirmed by oil red O staining as described previously (Fig. 1a; see **Note 1**) [8, 9, 15, 16].

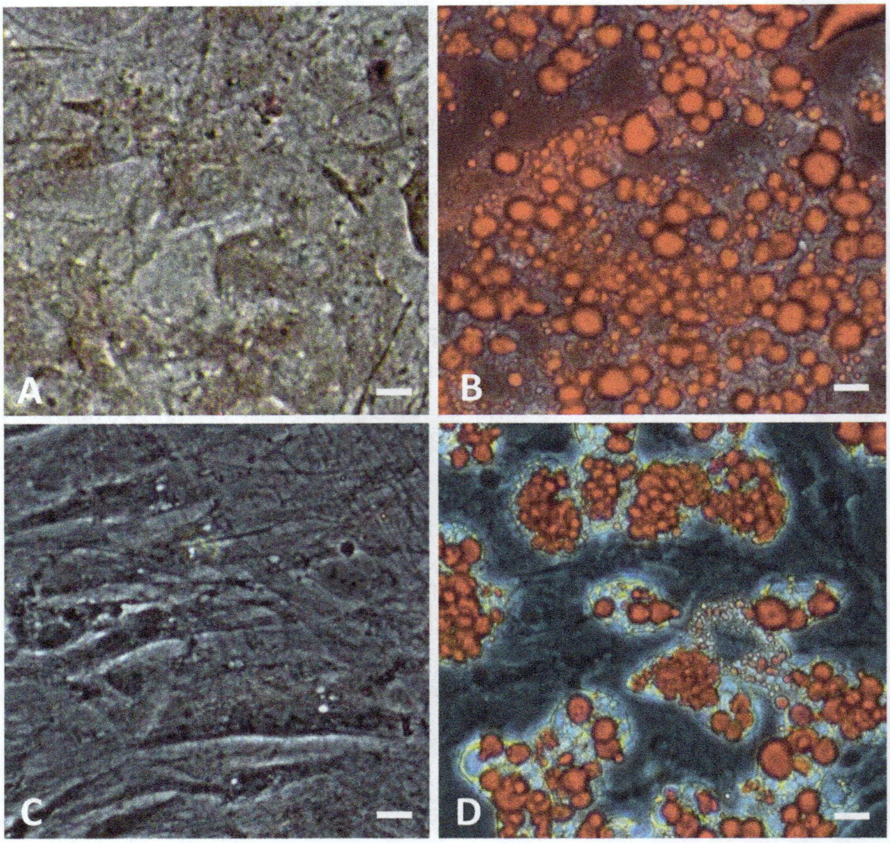

Fig. 1 Differentiation of 3T3-L1 cells and SVF cells. In the absence of differentiation induction (**a**), 3T3-L1 preadipocytes had marginal lipid accumulation on day 12. In the presence of differentiation induction (**b**), 3T3-L1 cell were fully differentiated into mature adipocytes and showed drastic fat accumulation on day 12 (**b**). Panels (**c**) (without differentiation induction) and (**d**) (with differentiation induction) showed that SVF cells had similar differentiation pattern with substantial lipid accumulation on day 10. Scale bar, 50 μm

3.2 Differentiation of SVF Cells

1. Seed the isolate SVF cells in 10-cm dishes (collagen-coated dishes preferred) then subculture in 6-well plates (either collagen-coated or regular tissue culture plates work well) with basal media.
2. Change the media every 2 days until the cells reach confluence (day 0).
3. Change the media one more time and maintain the cells in basal media till day 2.
4. At the end of day 2, replace basal media with differentiation medium; supply fresh differentiation medium every 2 days.
5. At the end of day 6, replace differentiation medium with maintenance medium. At this point, development of lipid droplets in the cells becomes obvious.
6. Supply fresh maintenance medium every 2 days till day 10, when adipocyte differentiation and lipid accumulation are complete, which is confirmed by oil red O staining as described previously (Fig. 1b; see **Note 2**) [14].

3.3 Kinetics of Autophagy Markers During Adipocyte Differentiation

1. At the indicated time points during cell differentiation (Fig. 2), remove the media and wash the cells with cold PBS twice.
2. Harvest the cells in 2 ml cold PBS using cell lifters.
3. Centrifuge at 5000 × *g*, 4 °C for 5 min to pellet the cells, and discard the supernatant.
4. Lyse the cells with a bullet blender according to the manufacturer's instruction.
5. Centrifuge at 12,000 × *g*, 4 °C for 10 min.
6. Transfer the cell lysate to a clean Eppendorf tubes, and measure total protein concentration on a microplate reader with DC protein assay kits.
7. Perform western blot analysis, using 12.5% gel for SDS-PAGE (60 V for 30 min, followed by 110 V for 2 h; see **Note 3**).
8. Use ECL kits and X-ray film for western blot detection of the steady-state level of autophagy proteins beclin 1, LC3-II, and p62 on day 0 through day 12 (Fig. 2; see **Note 4**).

3.4 Autophagy Flux in Preadipocytes and Mature Adipocytes

1. At day 12, 3T3-L1 preadipocytes (without differentiation induction) and mature adipocytes (with differentiation induction) in maintenance media are treated with bafilomycin A1 (inhibitor of autophagosome acidification, at 0.1 μM) plus leupeptin (the inhibitor of lysosomal proteases, at 10 μg/ml) for 4 h. Cells treated with DMSO serve as the controls (see **Notes 5** and **6**).
2. Remove the media and wash the cells with cold PBS twice.
3. Harvest the cells in 2 ml cold PBS using cell lifters.

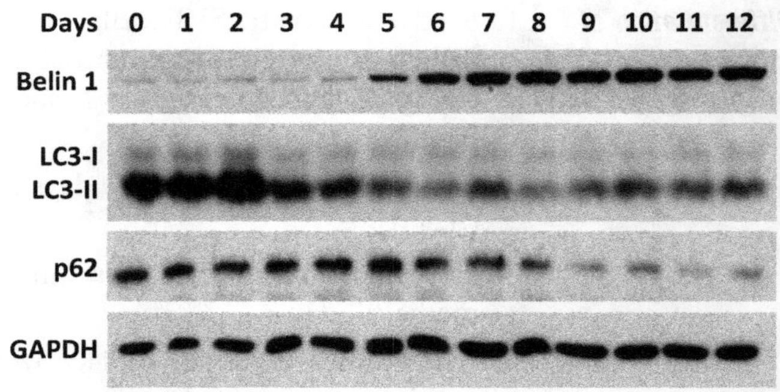

Fig. 2 The kinetics of autophagy protein levels during 3T3-L1 adipocyte differentiation. Beclin 1, the regulator of autophagy initiation, was upregulated. By contrast, LC3-II and p62, the proteins that can be selectively degraded by autolysosome, were downregulated. GAPDH served as the loading control. The time course of autophagy protein changes indicated activation of autophagy during adipocyte differentiation. However, note the limitation of using steady-state level of autophagy proteins to assess autophagy activity (see **Notes 4** and **7**)

4. Centrifuge at 5000 × *g*, 4 °C for 5 min to pellet the cells, and discard the supernatant.
5. Lyse the cells with a bullet blender according to the manufacturer's instruction.
6. Centrifuge at 12,000 × *g*, 4 °C for 10 min.
7. Transfer the cell lysate to a clean Eppendorf tube, and measure total protein concentration on a microplate reader with DC protein assay kits.
8. Proceed to western blot analysis to detect the protein levels of LC3-II and p62 in the absence and presence of autophagy inhibitors bafilomycin A1 plus leupeptin (B + L).
9. Use NIH ImageJ software to perform densitometry analyses of western blot images.
10. Autophagy flux is calculated as the difference of band density between B + L treated cells and untreated cells, for preadipocytes and differentiated (or mature) adipocytes, respectively (Fig. 3; see **Notes** 7 and **8**).

4 Notes

1. The expression of PPARγ (a key adipogenic regulator) and adiponectin (adipocytokine secreted by adipocytes) can be used as molecular markers to confirm adipocyte differentiation. Our experience indicates that after day 12 there is no further increase in lipid accumulation or expression PPARγ and adiponectin.

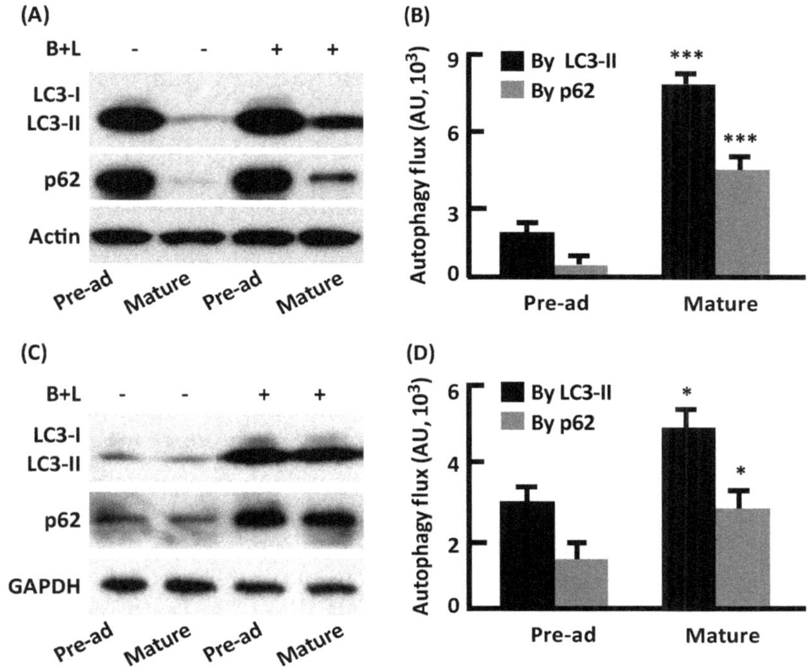

Fig. 3 Measurements of autophagy activity in preadipocytes and differentiated adipocytes (or mature adipocytes). (**a**) 3T3-L1 preadipocytes (pre-ad) and differentiated adipocytes were treated with bafilomycin A1 and leupeptin (i.e., B + L) for 4 h, and cell lysates were prepared to probe LC3-II and p62 by western blot analysis. (**b**) Autophagy flux was calculated, (band density of LC3-II with B + L)—(band density of LC3-II without B + L), for 3T3-L1 cells. (**c**) SVF cells and differentiated SVF cells were treated with bafilomycin A1 and leupeptin (i.e., B + L) for 4 h, and cell lysates were prepared to probe LC3-II and p62 by western blot analysis. (**d**) Autophagy flux was calculated, (band density of LC3-II with B + L)—(band density of LC3-II without B + L), for SVF cells. GAPDH and β-actin served as the loading controls. $n = 4$–6, $^{*}p < 0.05$; $^{***}p < 0.001$ (comparing preadipocytes with mature adipocytes)

2. Our experience indicates that after day 10 there is no further increase in lipid accumulation or expression of PPARγ and adiponectin.

3. SDS-PAGE at a lower voltage when necessary, e.g., 30 V for overnight (12–14 h) to increase the separation efficiency for LC3-I and LC3-II.

4. The steady-state level of beclin 1 increases over time during cell differentiation, indicative of an activated autophagy. However, the steady-state level of LC3-II is downregulated during cell differentiation, seemingly suggestive of inhibition of autophagy. This complexity arises from the dual role of LC3-II in autophagy: (a) promoter of phagophore formation and autophagosome maturation, and (b) autophagy substrate for degradation and recycling [12]. As such, the decreased steady-state level of LC3-II protein may be due to (1) elevated autophagy degradation, (2) reduced expression of LC3, or (3) compromised conversion of

LC3-I into LC3-II. In case (1), autophagy is upregulated, whereas in cases (2) and (3), autophagy is dampened. The decrease in p62 protein seems to support the case (1), i.e., activation of autophagy to degrade p62 and LC3-II. However, the conclusion is insecure before it is determined whether the expression of p62 and LC3 undergoes downregulation, and whether the conversion of LC3-I to LC3-II is compromised. Fortunately, the above-mentioned complexity and uncertainty can be solved by autophagy flux analysis.

5. For SVF cells, the B + L treatment starts on day 10. Preadipocytes (without differentiation induction) and mature adipocytes (with differentiation induction) in maintenance media are treated with bafilomycin A1 (0.1 μM) plus leupeptin at 10 μg/ml for 4 h. Cells treated with DMSO serve as the controls.
6. If it is desirable to measure autophagy flux on any specific day (e.g., day 6) during differentiation, the cells (with or without differentiation induction) will be treated with B + L on day 6 for 4 h, and then proceed to step 2 of Sect. 3.4.
7. The autophagy flux measurements based on LC3-II and p62 consistently suggest a higher substrate turnover (i.e., autophagy activity) in differentiated cells than in undifferentiated cells. Given the complexity discussed in **Note 4**, autophagy flux is superior to steady-state parameters in the assessment of autophagy activity.
8. Autophagy flux is upregulated in differentiated 3T3-L1 and SVF cells. However, autophagy is activated to a greater extent in 3T3-L1 cells than in SVF cells during differentiation (Fig. 3). This is in line with the lesser potential of differentiation known for SVF cells (Fig. 1), and with the notion that autophagy is required for adipocyte differentiation [8–11].

Acknowledgment

This work was supported in part by USDA National Institute of Food and Agriculture Hatch Project 1007334 (ZC).

References

1. Kershaw EE, Flier JS (2004) Adipose tissue as an endocrine organ. J Clin Endocrinol Metab 89:2548–2556
2. Galic S, Oakhill JS, Steinberg GR (2010) Adipose tissue as an endocrine organ. Mol Cell Endocrinol 316:129–139
3. Slutsky N, Vatarescu M, Haim Y et al (2016) Decreased adiponectin links elevated adipose tissue autophagy with adipocyte endocrine dysfunction in obesity. Int J Obes (Lond) 40:912–920
4. Rutkowski JM, Stern JH, Scherer PE (2015) The cell biology of fat expansion. J Cell Biol 208:501–512
5. Crewe C, An YA, Scherer PE (2017) The ominous triad of adipose tissue dysfunction: inflammation, fibrosis, and impaired angiogenesis. J Clin Invest 127:74–82

6. Mizushima N, Komatsu M (2011) Autophagy: renovation of cells and tissues. Cell 147:728–741
7. Green DR, Levine B (2014) To be or not to be? How selective autophagy and cell death govern cell fate. Cell 157:65–75
8. Liu L, Tao Z, Zheng LD et al (2016) FoxO1 interacts with transcription factor EB and differentially regulates mitochondrial uncoupling proteins via autophagy in adipocytes. Cell Death Dis 2:16066
9. Liu L, Zheng LD, Zou P et al (2016) FoxO1 antagonist suppresses autophagy and lipid droplet growth in adipocytes. Cell Cycle 15:2033–2041
10. Singh R, Xiang Y, Wang Y et al (2009) Autophagy regulates adipose mass and differentiation in mice. J Clin Invest 119:3329–3339
11. Zhang Y, Goldman S, Baerga R, Zhao Y, Komatsu M, Jin S (2009) Adipose-specific deletion of autophagy-related gene 7 (atg7) in mice reveals a role in adipogenesis. Proc Natl Acad Sci U S A 106:19860–19865
12. Mizushima N, Yoshimori T, Levine B (2010) Methods in mammalian autophagy research. Cell 140:313–326
13. Klionsky DJ, Abdelmohsen K, Abe A et al (2016) Guidelines for the use and interpretation of assays for monitoring autophagy (3rd edition). Autophagy 12:1–222
14. Liu L, Zheng LD, Donnelly SR, Emont MP, Wu J, Cheng Z (2017) Isolation of mouse stromal vascular cells for monolayer culture. Methods Mol Biol 1566:9–16
15. Liu L, Zou P, Zheng L et al (2015) Tamoxifen reduces fat mass by boosting reactive oxygen species. Cell Death Dis 6:e1586
16. Zou P, Liu L, Zheng L et al (2014) Targeting FoxO1 with AS1842856 suppresses adipogenesis. Cell Cycle 13:3759–3767

Methods in Molecular Biology (2018) 1854: 55–70
DOI 10.1007/7651_2017_66

Published online: 27 August 2017

Determination of Autophagy in the Caco-2 Spontaneously Differentiating Model of Intestinal Epithelial Cells

Sinem Tunçer and Sreeparna Banerjee

Abstract

The Caco-2 colorectal cancer cell line is widely used as a model for intestinal differentiation and barrier function. These cells, upon reaching confluency, spontaneously differentiate into enterocyte-like cells, synthesize intestinal enzymes, and form domes. Caco-2 cells also undergo autophagy in the course of differentiation. The criteria to establish the induction of autophagy in cells are already well established. Here, we describe the protocol for the spontaneous differentiation of Caco-2 cells and the detection of autophagy using Western blot, flow cytometry, and immunofluorescence.

Keywords Autophagy, Caco-2, Colon, Spontaneous differentiation

1 Introduction

Autophagy is a cellular mechanism for the physiological processes of degradation and recycling of organelles, long lived proteins, and other intracellular components [1]. Macroautophagy (referred to as autophagy), the most well studied of all autophagy processes, is considered as a cell survival mechanism as a response to stresses such as starvation whereby cellular materials are enclosed in double membrane layered organelles known as autophagosomes for degradation [2]. The formation of autophagosomes involves a signaling mechanism that is initiated with the uncoupling of the protein Beclin-1 from the mitochondrial membrane and the subsequent formation of a phagophore nucleation site where phospholipid effectors and autophagy-related proteins are recruited [3]. This is followed by the continual incorporation of membrane into the structure. The autophagy-related protein LC3 (consisting of LC3A, LC3B, and LC3C) is an important marker for autophagosome formation [4]. This ubiquitin-like protein undergoes a number of post-translational changes including lipidation, yielding LC3-II, which is then attached to the growing autophagosome membrane. LC-3 plays a number of roles in autophagy, including fusion of membrane components, selection of cargo and transport of the autophagosome [5]. LC3-II to LC3-I ratio is extensively used as a reliable index of autophagy [6, 7]. The autophagosome next fuses with the lysosome where further maturation and acidification takes place [1]. The packaging of cargo destined for degradation during autophagy is mediated by a protein called p62/SQSTM1, which is itself destroyed during autophagy. Therefore, the loss of p62 is considered as a marker for autophagic flux; accumulation of p62, on the other hand, is considered as an indicator of impaired autophagy [8]. However, the level of p62 can also be modulated independent of autophagy, therefore, additional methods are recommended to assess autophagic flux [9]. One key approach is to monitor LC3 turnover [10, 11]. Enhanced accumulation of autophagic vesicles should occur when lysosomal function is inhibited by chloroquine, a drug that impedes the fusion of autophagosomes and lysosomes and/or the activity of autolysosomes. Accordingly, in the presence of chloroquine, LC3-II accumulation would indicate efficient autophagic flux, while failure of LC3-II protein to increase would represent a defect or delay earlier in the autophagy process, prior to degradation at the autolysosome [12].

The colonic epithelium consists of a single layer of cells that has a high rate of turnover and therefore is in a state of continuous renewal. Multipotent stem cells exist in the intestinal crypts that continuously divide to generate transit amplifying cells, which then differentiate into specialized absorptive cells [13]. Being a catabolic process that responds to environmental cues, autophagy is activated during differentiation, most likely to orchestrate the rapid changes that are required [14]. In particular, autophagy is known to be essential for homeostasis in post-mitotic differentiated cells, primarily as a means for quality control of both proteins and organelles [14].

The Caco-2 cell line, originally isolated from a patient with colon adenocarcinoma, is widely used as a model for intestinal differentiation and barrier formation [15]. When grown to confluency, these cells show morphological as well as functional characteristics of terminally differentiated enterocytes such as presence of brush borders on the apical surface, enhanced tight junctions, dome formation, and expression of enzymes such as sucrase isomaltase [15, 16]. Using this cell line as a model of terminal intestinal differentiation that can be easily maintained and manipulated, we report here detailed protocols to establish and confirm spontaneous differentiation of Caco-2 cells. Additionally, we report protocols for the identification of autophagy in the differentiated cells by enhanced expression of the autophagy marker Beclin-1, lipidation of LC-3 (LC3-II/LC3-I ratio), and determination of autophagic puncta formation using the cationic tracer dye CYTO-ID® by immunofluorescence and flow cytometry. In addition, the presence of autophagic flux was established by determining the levels of p62 and LC-3 turnover in the presence of the lysosomotropic reagent chloroquine.

2 Materials

2.1 Reagents and Materials for Cell Culture and Differentiation

Lists of reagents and materials for the propagation, subcultivation, cryopreservation, and differentiation of Caco-2 cells are given. All disposable plasticware used in this study were supplied by Greiner Bio-One (Greiner Bio-One, Kremsmünster, Austria) (*see* **Note 1**).

2.1.1 Complete Growth Medium (Earl's Minimal Essential Medium) Components

1. 20% Fetal Bovine Serum (FBS) (Biological Industries, Beit Haemek, Israel, cat no: 04-127-1A).
2. 100 U–100 mg Penicillin-Streptomycin (Biological Industries, cat no: 03-031-1B).
3. 2 mM L-glutamine (Biological Industries, cat no: 03-020-1B).
4. 1× NEA-Non-Essential Amino Acids (Biochrom GmbH, Berlin, Germany, cat no: K 0293).
5. 1 mM Na pyruvate (Biowest, Florida, USA, cat no: L0642-100).
6. EMEM-Minimum Essential Medium, no glutamine, no phenol red, with Earl's salt (Thermo Fisher Scientific, Boston, MA, USA, cat no: 51200038).

2.1.2 Sub-culturing Reagents

1. PBS (Dulbecco's Phosphate Buffered Saline, without Ca^{2+} and Mg^{2+}) (Biological Industries, cat no: 02-023-1A).
2. Trypsin/EDTA solution, 0.25%/0.02% w/v (Biochrom GmbH, cat no: L2163).
3. Complete growth medium.
4. 37 °C water bath.

2.1.3 Materials for Cryopreservation

1. Freezing medium: 5.0% Dimethyl Sulfoxide (DMSO) (PanReac AppliChem, Darmstadt, Germany, cat no: A7248, 0010) containing complete growth medium.
2. Mr. Frosty™ Freezing Container (Thermo Fisher Scientific, cat no: 5100-0001).
3. Liquid nitrogen.

2.2 Reagents and Materials for RNA Isolation, cDNA Synthesis, and Quantitative Real-Time PCR (qRT-PCR)

2.2.1 RNA Isolation

1. PBS.
2. RNeasy Mini kit (Qiagen, Hilden, Germany, cat no: 74106).
3. BioDrop μLITE spectrophotometer (BioDrop, Cambridge, UK).

2.2.2 cDNA Synthesis and qRT-PCR

1. RNase-free DNase I (1 U/μL), supplied with $MnCl_2$ (Thermo Fisher Scientific, cat no: EN0521).
2. RevertAid First Strand cDNA Synthesis Kit (Thermo Fisher Scientific, cat no: K1622).
3. Thermal cycler.
4. FastStart Universal SYBR Green Master (Rox) (Roche, Basel, Switzerland, cat no: 04913850001).
5. Rotor-Gene Q6000 real-time PCR machine (Qiagen).
6. Primers (Table 1).

2.3 Reagents and Materials for Protein Isolation and Western Blot

2.3.1 Protein Isolation

1. PBS.
2. M-PER Mammalian Protein Extraction Reagent (Thermo Fisher Scientific, cat no: 78501) containing Mini Protease Inhibitor Cocktail, cOmplete™ (Roche, cat no: 11836153001), and inhibitor tablets for phosphatase, Phos-STOP™ (Roche, cat no: 04906837001).
3. Cooling centrifuge.
4. Coomassie (Bradford) Protein Assay Kit (Thermo Fisher Scientific, cat no: 23200).
5. Sample loading dye (6×): 12% SDS, 30% β-mercaptaethanol, 30% glycerol, 0.02% bromophenol blue, 375 mM Tris–HCl pH 6.8. Aliquot and store at −20 °C.
6. 95 °C heat block.

Table 1
Sequences of primers used in qRT-PCR

Name	Primer sequence (5′ → 3′)	Annealing temperature (°C)	Product size (bp)	Target mRNA ID
Sucrase isomaltase (SI) mRNA	Forward: CAAATGGCCAAACACCAATG	60	160	NM_001041.3
	Reverse: CCACCACTCTGCTGTGGAAG			
β-Actin mRNA	Forward: CAGCCATGTACGTTGCTATCCAGG	60	151	NM_003900.4
	Reverse: AGGTCCAGACGCAGGATGGCATG			

Primer sequences, amplified product sizes, and target mRNA IDs (GRCh38.p7 Primary Assembly, NCBI) are given

2.3.2 SDS Gel Electrophoresis, Transfer and Immunoblotting

1. 4% Stacking gel: For 1.5 mm spacer, prepare 4.0 mL of stacking gel as 2.52 mL dH_2O, 300 μL 40% acrylamide/bisacrylamide solution (29:1), 1.0 mL 0.5 M Tris–HCl pH 6.8, 40 μL 10% SDS, 4.0 μL TEMED, 20 μL 10% APS.
2. 12% Separating gel: Prepare 10 mL of separating gel for 1.5 mm spacer as 4.33 mL dH_2O, 3.00 mL 40% acrylamide/bisacrylamide solution (29:1), 2.5 mL 1.5 M Tris–HCl pH 8.8, 100 μL 10% SDS, 5.0 μL TEMED, 50 μL 10% APS.
3. PageRuler Plus Prestained Protein Ladder (Thermo Fisher Scientific, cat no: 26619).
4. SDS-PAGE running buffer (10×): 250 mM Tris base, 1.92 M glycine, 1.0% SDS. Store the buffer at room temperature. For 1× SDS-PAGE running buffer, dilute 100 mL of 10× SDS-PAGE running buffer with 900 mL of dH_2O.
5. Mini-PROTEAN® Tetra Vertical Electrophoresis Cell for Mini Precast Gels (Bio-Rad, cat no: 1658005) for SDS gel electrophoresis.
6. Polyvinylidene Difluoride (PVDF) membrane with a pore size of 0.2 μm (Roche, cat no: 03010040001).
7. Transfer Buffer (10×): 250 mM Tris base, 1.92 M glycine. Store transfer buffer (10×) at 4 °C. For 1× transfer buffer, mix 100 mL 10× transfer buffer, 200 mL methanol and 700 mL dH_2O and prechill at 4 °C.
8. Mini Trans-Blot® Module (Bio-Rad, cat no: 1703935) for blotting.
9. Rotilabo® Blotting paper, 0.35 mm thick (Carl Roth GmbH, Karlsruhe, Germany, cat no: CL67.1).
10. PowerPac Basic Power Supply (Bio-Rad, cat no: 164-5050).
11. Tris Buffered Saline (TBS, 10×): 250 mM Tris base, 1.5 M NaCl, 20 mM KCl. Adjust pH to 7.4 ± 0.2. Sterilize by filtration or autoclaving, and store at room temperature. For a 1× solution, mix 1 part of the 10× solution with 9 parts dH_2O. For 1× TBS with 0.1% Tween 20 (0.1% TBS-T), add 0.1% Tween 20 in 1× TBS.
12. Blocking solution: 5% skim milk (Sigma-Aldrich, St. Louis, MO, USA, cat no 70166) in 0.1% TBS-T. Store at 4 °C.
13. Antibody diluent solutions: 5% skim milk (Sigma-Aldrich) or 5% Albumin Fraction V (PanReac AppliChem, cat no A1391) in 0.1% TBS-T. Store at 4 °C.
14. Primary and secondary antibodies (Table 2).
15. Chemiluminescent Clarity ECL Substrate (Bio-Rad, cat no: 1705060) for detection of horse radish peroxidase (HRP).

16. ChemiDoc MP Imaging System (Bio-Rad) for chemiluminescent detection.
17. Membrane stripping reagents: Dissolve 15 g glycine, 1 g SDS, and 10 mL Tween 20 in 800 mL dH2o. Adjust pH lo 2.2, and bring volume up to 1 L with dH2O.

2.4 Reagents and Materials for Flow Cytometry

1. Phenol red free complete cell growth medium containing 5.0% FBS.
2. CYTO-ID® Autophagy detection kit (Enzo Life Sciences, Farmingdale, NY, USA, cat no: ENZ-51031-0050).
3. BD Accuri™ C6 flow cytometer (BD Biosciences, Ann Arbor, MI, USA).
4. 37 °C heat block.

2.5 Reagents and Materials for Confocal Microscopy

1. Phenol red free complete cell growth medium containing 5.0% FBS.
2. Glass bottom dish: 35 mm Glass bottom dish with 20 mm micro-well #0 cover glass (Cellvis, Mountain View, CA, USA, cat no: D35-20-0-N).
3. CYTO-ID® Autophagy detection kit (Enzo Life Sciences).
4. 37 °C incubator.

Table 2
Primary and secondary antibodies used for Western blot

Primary antibodies and dilutions	Source of primary antibodies	Secondary antibodies and dilutions	Source of secondary antibodies
LC3A/B (D3U4C) XP® 1:1000 in 5% BSA in TBS-T	Cell signaling Technology (Leiden, The Netherlands) cat no: 12741	Goat anti-rabbit IgG-HRP 1:2000 in 5% skim milk TBS-T	Santa Cruz Biotechnology cat no: sc-2004
Beclin-1 (D40C5) 1:1000 in 5% BSA in TBS-T	Cell signaling Technology, cat no: 3495	Goat anti-rabbit IgG-HRP 1:2000 in 5% skim milk TBS-T	Santa Cruz Biotechnology, cat no: sc-2004
SQSTM1/p62 (M01), clone 2C11 1:1000 in 5% skim milk in TBS-T	Abnova (Taipei, Taiwan), cat no: H00008878-M01	Goat anti-mouse IgG-HRP 1:2000 in 5% skim milk TBS-T	Santa Cruz Biotechnology, cat no: sc-2005
β-Actin (C4) 1:4000 in 5% skim milk in TBS-T	Santa Cruz Biotechnology, (Santa Cruz, CA, USA) cat no: sc-47778	Goat anti-mouse IgG-HRP 1:2000 in 5% skim milk TBS-T	Santa Cruz Biotechnology, cat no: sc-2005

A detailed inventory of each primary antibody, LC3-I/II (14/16 kDa), Beclin-1 (60 kDa), p62/SQSTM1 (62 kDa), β-Actin (42 kDa), and each corresponding secondary antibody used in this study are given, including working dilutions and supporting documentation

5. Formaldehyde solution, 37% for molecular biology (Panreac AppliChem, cat no: A0877). For 4% formaldeheyde, dilute 37% formaldehyde in PBS.
6. Confocal microscope with a standard FITC filter set. In the study, Leica DMI4000 with Andor DSD2 Spinning Disc Confocal Microscope equipped with a Leica 63×/1.40 Oil HC PL APO objective (Leica Microsystems, Wetzlar, Germany) was utilized to monitor cells. GFP (Green Fluorescence Protein) filter set was used for imaging.

2.6 Statistical Analyses

Results were expressed as the mean ± standard error of mean (SEM). Graphpad (Prism version 6.00 for Windows, GraphPad Software, La Jolla, California USA) was used for data analyses and graphing. Statistical analysis was carried out using unpaired *t*-test.

3 Methods

3.1 Cell Culture and Spontaneous Differentiation of Caco2 Cells

3.1.1 Cell Thawing

1. Add 9 mL of complete growth medium to a 15 mL conical centrifuge tube.
2. Thaw a vial of frozen cells in a water bath set at 37 °C. Thawing should be rapid (approximately 1 min), just long enough for most of the ice to melt.
3. Remove vial from the water bath and process immediately. Transfer the contents of the vial to a 15 mL conical centrifuge tube containing 9 mL complete growth medium, slowly.
4. Centrifuge at 100 × *g*, at room temperature, for 15 min.
5. Carefully aspirate the medium.
6. Re-suspend the pellet gently by pipetting up and down in 5 mL of complete growth medium, and dispense into a 25 cm^2 culture flask.

3.1.2 Cell Line Maintenance and Sub-culturing

Volumes are given for a 25 cm^2 flask. Adjust the amounts proportionally for culture vessels of other sizes.

When the cells are about 50–60% confluent in a 25 cm^2 flask (*see* **Note 2**),

1. Remove the medium from the dish and briefly rinse the cells with PBS to remove all traces of medium containing FBS.
2. Discard the PBS, and add 1 mL of trypsin-EDTA solution to the flask and incubate the cells at 37 °C incubator.
3. Observe the cells under an inverted microscope until cell layer is dispersed which is usually within 5–10 min (*see* **Note 3**).
4. Add 3.0 mL of complete growth medium and disperse the cells by gently pipetting (*see* **Note 4**).

5. Add appropriate aliquots of the cell suspension to new culture vessels. A subcultivation ratio of 1:3 is recommended.
6. Incubate cultures at 37 °C in a humidified, 5% CO_2 incubator.
7. Change the medium every 2 days.

3.1.3 Cell Freezing

Sub-confluent Caco-2 cells (50–60% confluent in a 25 cm^2 flask) are detached using the standard procedure described in Sect. 3.1.2. Prior to detaching cells, prepare the freezing medium and keep on ice.

1. After centrifugation at 100 × *g*, for 5 min at room temperature, remove the medium, and resuspend the cells in 2 mL of ice-cold freezing medium.
2. Distribute to two cryotubes and keep at −80 °C in a Mr. Frosty™ Freezing Container overnight, then transfer the cells to the vapor phase of a liquid nitrogen freezer.

3.1.4 Protocol for Spontaneous Differentiation of Caco-2 Cells

Differentiation can be carried out in 25 cm^2 flask (Greiner Bio-One) (*see* **Note 1**), as well as 6- or 12-well plates (Greiner Bio-One).

1. Seed the cells into culture dishes, plates or flasks at a density of approximately $3.5 \times 10^4/cm^2$.
2. Two days post seeding, the cells should reach 100% confluency (day 0).
3. Change the medium every other day after the cells reach confluency. Post-confluent Caco-2 cells cultured for 10–20 days are considered to be differentiated. Dome formation, as a differentiation phenotype, can be easily observed with an inverted light microscope (Fig. 1).

3.2 Detection of Caco-2 Cell Differentiation via qRT-PCR for Sucrase Isomaltase

Increase in the expression of sucrase isomaltase (SI) is considered to be a marker of Caco-2 differentiation [17, 18]. qRT-PCR can be used to determine SI expression during the differentiation process.

3.2.1 Total RNA Isolation and cDNA Synthesis

1. Collect the cells with a scraper on ice, after washing with ice-cold PBS.
2. Isolate total RNA, and determine quantity and quality of the isolated RNA.
3. To remove traces of genomic DNA before cDNA synthesis, carry out a DNAse I treatment. In this study, 0.5 μg of total RNA was treated with 1 U RNase-free DNase I in the presence of $MgCl_2$ containing 10× DNAse buffer for 30 min at 37 °C. DNase treatment was stopped by incubation at 65 °C for 10 min, after addition of 1 μL of 50 mM EDTA.

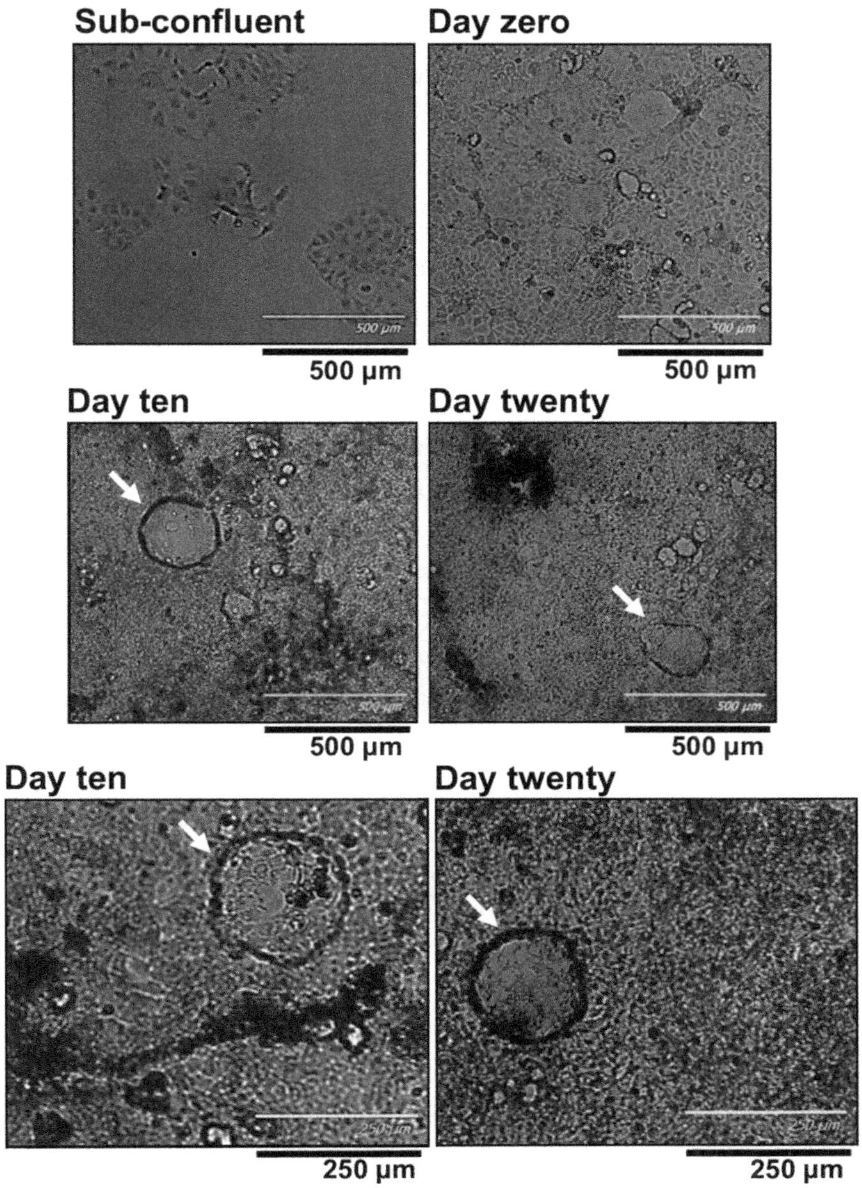

Fig. 1 Caco-2 cells form domes during spontaneous differentiation. Caco-2 cells were assessed using inverted light microscopy for morphological changes during spontaneous differentiation. Representative images are given for sub-confluent cells, confluent (day 0) cells, and 10 and 20 days differentiated cells. Domes are indicated with arrows

4. Carry out first strand cDNA synthesis using reverse transcriptase. In this study, first strand cDNA synthesis with random primers was done using RevertAid First Strand cDNA Synthesis Kit, according to the manufacturer's instructions.

3.2.2 qRT-PCR

Expression level of SI was determined using quantitative real-time polymerase chain reaction (qRT-PCR). *ACTB* was used as endogenous reference gene [19]. Primer sequences and annealing temperatures for *ACTB* and *SI* are given in Table 1.

1. Generate standard curves for both *ACTB* and *SI* by plotting the log of the serial dilutions of cDNA template against the cycle threshold (Ct) values obtained during amplification of each dilution. Calculate amplification efficiency from the slope of the standard curve.
2. Dilute sample cDNA as 1:10 with nuclease free dH_2O.
3. For a 20 μL reaction, prepare the reaction as:

   2.0 μL 1:10 diluted cDNA (*see* **Note 5**).
   0.25 μM of each primer (0.5 μL each from 10 μM stock).
   10 μL SYBR Green Master Mix.
   7.0 μL nuclease free dH_2O.
4. Follow the cycling conditions as 95 °C for 10 min; three step cycling for 40 times (95 °C for 30 s, 60 °C for 30 s and 72 °C for 30 s); and MCA (Melting Curve Analysis) settings of a ramp from 50 to 99 °C, rising by 1.0 °C in each step.

   In this study, relative expression of SI was evaluated by the Pfaffl method [20]. The results were normalized with respect to SI expression in confluent (day 0) cells, and represented as "fold increase" (Fig. 2) (*see* **Note 6**).

3.3 Autophagy Detection During Spontaneous Differentiation of Caco-2 Cells by Western Blot

3.3.1 LC3 and Beclin-1 Expression

Western blot was used for the detection of LC3-I, LC3-II (lipidated), and Beclin-1 during spontaneous differentiation of Caco-2 cells (Fig. 3) (*see* **Note 7**).

1. After washing the cells with ice-cold PBS, add lysis buffer containing protease and phosphatase inhibitors. Collect the cells by scraping, and transfer to a 1.5 mL Eppendorf tube.
2. Vortex vigorously and incubate on ice for 30 min. During the incubation, vortex every 5 min.
3. Centrifuge at 14,000 × *g* for 10 min at 4 °C. Collect the supernatant in a fresh tube.
4. Determine protein concentrations by Coomassie Protein Assay Reagent relative to a standard curve generated with bovine serum albumin.
5. Mix the protein samples with 6× loading buffer and heat at 95 °C for 6 min in a heat block.

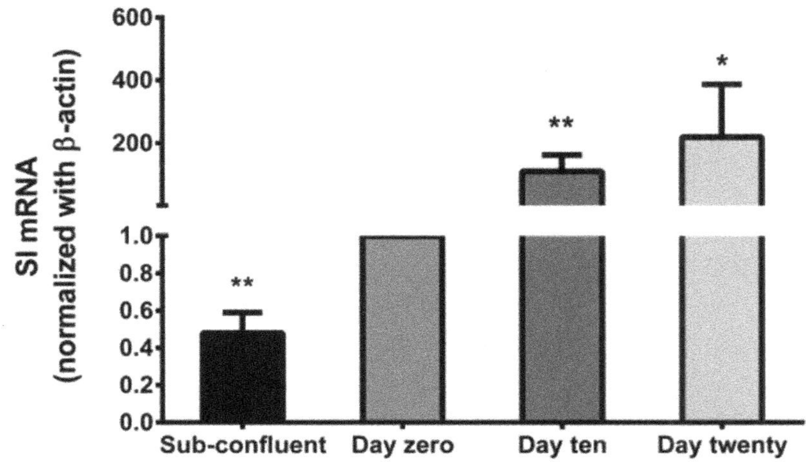

Fig. 2 Confirmation of differentiation in spontaneously differentiating Caco-2 by increased expression of sucrase isomaltase (SI). As a marker of enterocyte differentiation, sucrase isomaltase expression during differentiation was analyzed by qRT-PCR. Fold-changes in gene expression, calculated from four independent experiments with three technical replicates each, are given relative to confluent (day 0) cells. Data are represented as (mean ± SEM, $n = 4$). *t*-test was used for statistical analyses (*$p \leq 0.5$, **$p \leq 0.01$)

6. Separate 20–50 μg of proteins in 12% SDS-PAGE gel.
7. Transfer the proteins to PVDF membranes using standard techniques.
8. At the end of transfer, block the membrane in 5% skim milk in TBS-T.
9. After blocking, incubate the membrane with primary antibody overnight at 4 °C.
10. Wash 3× in 0.1% TBS-T, incubate the membrane for 1 h with HRP-conjugated secondary antibody at room temperature.
11. Wash the membrane 3× with 0.1% TBS-T, and visualize the bands using HRP substrate.
12. Determine band intensities.
13. Strip the membrane at 70 °C, for 3 min. After stripping, wash the membrane 3 X with 0.1% TBS-T at room temperature. The stripped blot can be used to detect p62 levels.

3.3.2 Determination of LC3 Turnover and p62 to Measure Autophagic Flux

To assess autophagic flux in the Caco-2 differentiation model, confluent (day 0) and differentiated (day 10) cells were treated with 30 μM of chloroquine (from a 60 mM stock prepared in deionized water) for 18 h. The differences in the protein levels of lipidated LC3-II in the presence and absence of chloroquine, which represent the amount of LC3 delivered to lysosomes for processing, were assessed by Western blot as described in Sect. 3.3.1 (Fig. 3) (*see* **Note 8**). Additionally, p62 degradation indicative of autophagic flux was determined by Western blot (Fig. 3).

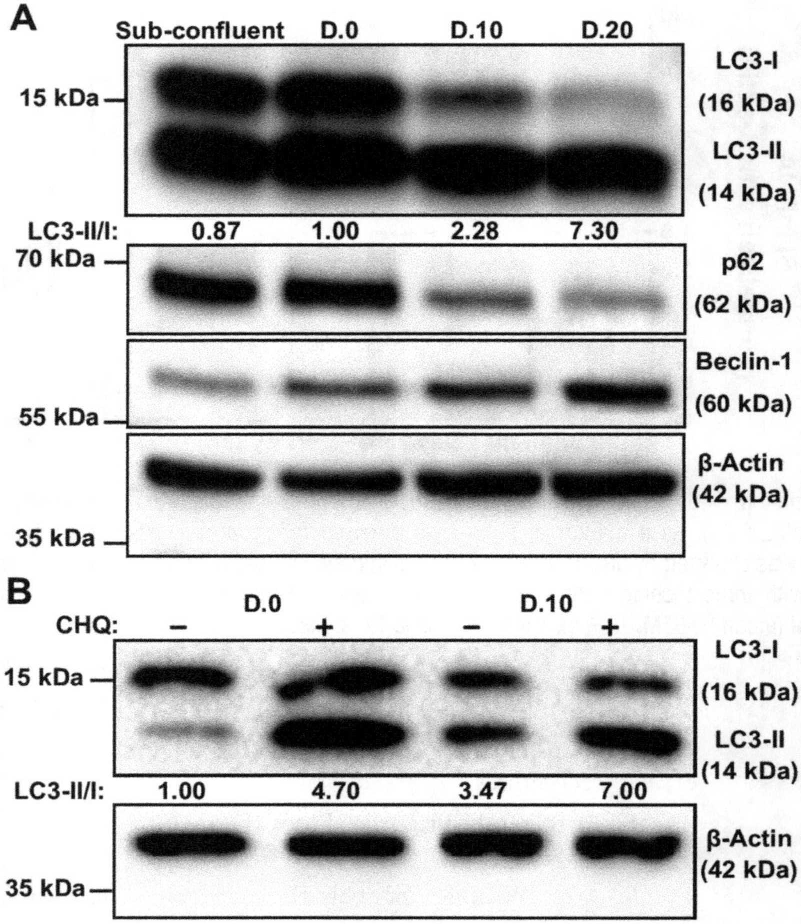

Fig. 3 Expression of autophagy-related proteins, and analysis of autophagic flux in undifferentiated and differentiated Caco-2 cells. (**a**) The expression levels of three autophagy-related proteins, LC3, Beclin-1, and p62, are shown. Ratio of LC3-II to LC3-I, as well as Beclin-1 expression increases while p62 level decreases in differentiated cells. (**b**) In the presence of chloroquine (CHQ), increase in the LC3-II to LC3-I ratio was seen in both undifferentiated (day 0) and differentiated (day 10) cells as an indication of efficient autophagic flux in differentiated Caco-2 cells

3.4 Monitoring Autophagic Activity in Spontaneously Differentiating Caco-2 Cells by Confocal Microscopy and Flow Cytometry

CYTO-ID® Autophagy Detection Kit was used to detect autophagic activity at the cellular level using microscopy and flow cytometry. CYTO-ID® Green autophagy dye specifically stains autophagosomes in live cells. 40–50% confluent Caco-2 cells were treated with 500 nM rapamycin (from a 500 μM stock in DMSO), a well-known inducer of autophagy, and 30 μM chloroquine, a lysosomal inhibitor, were used as positive control. Negative control included cells that were incubated in the complete culture media with vehicle (DMSO and deionized water for rapamycin and chloroquine, respectively) for an equal length of time under the same conditions.

3.4.1 Live Cell Analysis of Autophagy by Flow Cytometry

1. Differentiate Caco-2 cells in a 12-well plate for 10 days as described in Sect. 3.1.4.
2. Detach the cells by trypsinization (*see* **Note 9**), and pellet 2×10^5 cells (*see* **Note 10**) by centrifugation at $100 \times g$ for 5 min.
3. Wash the pellets in 1× Assay Buffer at $100 \times g$ for 5 min.
4. Prior to staining, dilute 1 μL of CYTO-ID® Green Detection Reagent in 1 mL of phenol red free complete cell growth medium containing 5% FBS.
5. Re-suspend the pellets in 250 μL of the diluted CYTO-ID® Green stain solution, and incubate for 30 min at 37 °C in the dark (*see* **Note 11**).
6. At the end of incubation, collect the cells by centrifugation at $100 \times g$ for 5 min.
7. Wash with 1× Assay Buffer at $100 \times g$ for 5 min.
8. Re-suspend the pellets in 500 μL of fresh 1× Assay Buffer, and analyze immediately in green channel (FL1) of a flow cytometry. Use unstained cells to determine the level of background fluorescence (Fig. 4).

3.4.2 Live Cell Analysis of Autophagy by Confocal Microscopy

1. Seed Caco-2 cells into 35 mm glass bottom dishes and differentiate until day 10, as described in Sect. 3.1.4 (*see* **Note 12**).
2. Wash sub-confluent, confluent (day 0) and 10 days differentiated Caco-2 cells with 1× Assay buffer.
3. Just before staining, dilute 2 μL of CYTO-ID® Green Detection Reagent in 1 mL of complete cell growth medium (without phenol red) containing 5% FBS, and dispense to glass as 100 μL.
4. Incubate the cells at 37 °C for 30 min, protected from light.
5. At the end of incubation, remove the detection reagent and wash the samples with 1× Assay Buffer.
6. Fix the cells with 4% formaldehyde in PBS for 20 min at room temperature.
7. Remove formaldehyde solution and wash the cells 3× with 1× Assay Buffer.
8. Analyze the stained cells by wide-field fluorescence or confocal microscope, with the FITC filter (Fig. 4).

4 Notes

1. Use of some other brands may cause cell detachment from the surface during spontaneous differentiation process.

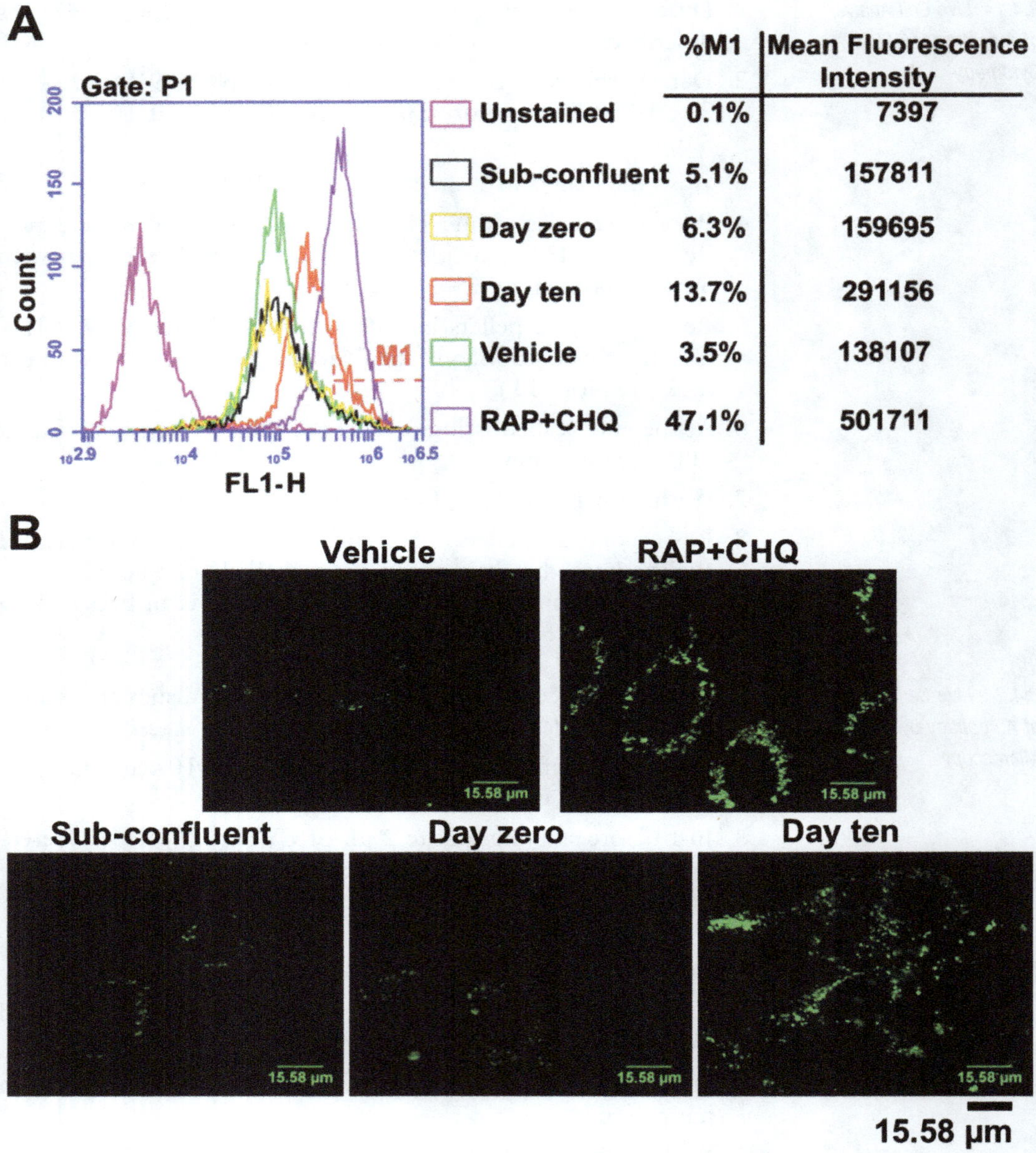

Fig. 4 Flow cytometry and confocal microscopy based profiling of autophagy in differentiating Caco-2 cells. In the experiments, rapamycin (RAP) and chloroquine (CHQ) treated sub-confluent cells were used as positive control. (**a**) Sub-confluent cells, confluent (day 0) cells, and 10 days differentiated cells were stained with CYTO-ID® Green Detection Reagent and analyzed by flow cytometry. Results are presented as histogram overlay. Increase in the magnitude of fluorescence signal (Mean Fluorescence Intensity), together with increased percentages of CYTO-ID® positive cell populations (M1%) was observed in differentiated cells. Sub-confluent cells were used for gating (M1), and unstained cells to determine the level of background fluorescence. (**b**) Fluorescent spots inside the Caco-2 cells represent CYTO-ID® Green autophagy dye stained autophagosomes/autolysosomes. Autophagy induction can be seen in differentiated cells

2. Since Caco-2 cells differentiate spontaneously, depending on cell density, always keep the cell density low to obtain a sub-confluent mono-layer (approximately 50–60% confluent).
3. Do not agitate the cells by hitting or shaking the flask to detach prematurely, as this may result in cell clumping.
4. Gently pipette the cell suspension up and down to dislodge and re-suspend the cells. Make sure you disperse the cell clumps into a single cell suspension. Cell clumps can lead to uneven growth and more clumping. To distribute cells, gently swirl the plate after cell seeding. Make sure that the cells are evenly dispersed in the wells or flask; otherwise, non-uniform distribution or an excessive amount of cell clumps may result in spontaneous differentiation, locally.
5. For standard curve generation, make serial dilutions of cDNA.
6. Although *SI* mRNA levels robustly increase during the differentiation process, increase in the expression levels may vary between 80- and 500-folds between the experimental sets.
7. Although the molecular weight of LC3-II is higher than LC3-I, it is observed to migrate faster by about 2 kDa than LC3-I in an SDS-PAGE. This is due to a high hydrophobicity of LC3-II rather than any post-translational processing [21].
8. CYTO-ID® Autophagy Detection Kit can also be used to detect autophagic flux. Incubation of the cells with chloroquine will increase the fluorescence signal with respect to vehicle treated controls, which is an indication of flux.
9. Incubation time with trypsin-EDTA should be enough to achieve a mono-dispersed cell suspension which increases staining efficiency.
10. Higher cell numbers result in inadequate staining.
11. During incubation, mix the cells gently by pipetting a few times to achieve a mono-disperse cell suspension.
12. After 10 days of incubation, Caco-2 cells were observed to detach from the glass surface of the dish.

Acknowledgements

This work was funded from TÜBİTAK Project no. 114S937 to S.B. Aslı Sade Memişoğlu, Orkun Cevheroglu, and members of the Banerjee lab are acknowledged for useful discussions. Cagdas Son and Mayda Gursel are acknowledged for sharing resources.

References

1. Tsapras P, Nezis IP (2017) Caspase involvement in autophagy. Cell Death Differ 1–11. doi:10.1038/cdd.2017.43
2. Klionsky DJ (2000) Autophagy as a regulated pathway of cellular degradation. Science 290:1717–1721. doi:10.1126/science.290.5497.1717
3. Kang R, Zeh HJ, Lotze MT et al (2011) The Beclin 1 network regulates autophagy and apoptosis. Cell Death Differ 18:571–580. doi:10.1038/cdd.2010.191
4. Wild P, McEwan DG, Dikic I (2014) The LC3 interactome at a glance. J Cell Sci 127:3–9. doi:10.1242/jcs.140426
5. Slobodkin MR, Elazar Z (2013) The Atg8 family: multifunctional ubiquitin-like key regulators of autophagy. Essays Biochem 55:51–64. doi:10.1042/bse0550051
6. Kadowaki M, Karim MR (2009) Cytosolic LC3 ratio as a quantitative index of macroautophagy. Methods Enzymol 451:199–213. doi:10.1016/S0076-6879(08)03613-6
7. Karim MR, Kanazawa T, Daigaku Y et al (2007) Cytosolic LC3 ratio as a sensitive index of macroautophagy in isolated rat hepatocytes and H4-II-E cells. Autophagy 3:553–560. doi:10.4161/auto.4615
8. Katsuragi Y, Ichimura Y, Komatsu M (2015) P62/SQSTM1 functions as a signaling hub and an autophagy adaptor. FEBS J 282:4672–4678. doi:10.1111/febs.13540
9. Klionsky DJ, Abdalla FC, Abeliovich H et al (2012) Guidelines for the use and interpretation of assays for monitoring autophagy. Autophagy 8:445–544. doi:10.4161/auto.19496
10. Zhang X, Chen S, Huang K et al (2013) Why should autophagic flux be assessed? Acta Pharmacol Sin 34:595–599. doi:10.1038/aps.2012.184
11. Mizushima N, Yoshimori T, Levine B (2010) Methods in mammalian autophagy research. Cell 140:313–326. doi:10.1016/j.cell.2010.01.028
12. Guo S, Liang Y, Murphy SF et al (2015) A rapid and high content assay that measures cyto-ID-stained autophagic compartments and estimates autophagy flux with potential clinical applications. Autophagy 11:560–572. doi:10.1080/15548627.2015.1017181
13. Simons BD, Clevers H (2011) Stem cell self-renewal in intestinal crypt. Exp Cell Res 317:2719–2724. doi:10.1016/j.yexcr.2011.07.010
14. Mizushima N, Levine B (2010) Autophagy in mammalian development and differentiation. Nat Cell Biol 12:823–830. doi:10.1038/ncb0910-823
15. Sambuy Y, De Angelis I, Ranaldi G et al (2005) The Caco-2 cell line as a model of the intestinal barrier: influence of cell and culture-related factors on Caco-2 cell functional characteristics. Cell Biol Toxicol 21:1–26. doi:10.1007/s10565-005-0085-6
16. Sun H, Chow EC, Liu S et al (2008) The Caco-2 cell monolayer: usefulness and limitations. Expert Opin Drug Metab Toxicol 4:395–411. doi:10.1517/17425255.4.4.395
17. Van Beers EH, Al RH, Rings EH (1995) Lactase and sucrase-isomaltase gene expression during Caco-2 cell differentiation. Biochem J 308:769–775. doi:10.1042/bj3080769
18. Buhrke T, Lengler I, Lampen A (2011) Analysis of proteomic changes induced upon cellular differentiation of the human intestinal cell line Caco-2. Dev Growth Differ 53:411–426. doi:10.1111/j.1440-169X.2011.01258.x
19. Dydensborg AB, Herring E, Auclair J et al (2005) Normalizing genes for quantitative RT-PCR in differentiating human intestinal epithelial cells and adenocarcinomas of the colon. Am J Physiol Gastrointest Liver Physiol 290:G1067–G1074
20. Pfaffl MW (2001)A new mathematical model-for relative quantification in real-time RT-PCR. Nucleic Acids Res 29:45e
21. Mizushima N, Yoshimori T, Levine B (2010) Methods in mammalian autophagy research. Cell 140:313–326. doi:10.1016/j.cell.2010.01.028

Methods in Molecular Biology (2018) 1854: 71–85
DOI 10.1007/7651_2018_123

Published online: 27 February 2018

Assessing Autophagy in the Leydig Cells

Hui Gao, Chao Liu, and Wei Li

Abstract

Autophagy is an important intracellular degradation system which is implicated in many physiological and pathological processes. During autophagy, cytosolic constituents such as organelles and macromolecules are engulfed by autophagosome, and then they fuse with lysosomes for degradation and recycle of the engulfed components within the autolysosome to maintain cellular homeostasis. In male testis, the Leydig cells provide the major source of testosterone production. Autophagy is extremely active in Leydig cells and is involved in the steroid production. However, the precise role of autophagy in Leydig cells is still largely unknown. Thus, a comprehensive measurement of autophagic activity with different methods would shed light on our knowledge about the functional role of autophagy in regulating male reproductive physiology. In this chapter, we describe the morphological, cellular, and biochemical methods to monitor autophagy in Leydig cells.

Keywords Autophagy, Immunoblotting, Immunofluorescence, LC3, Leydig cell, Transmission electron microscopy

1 Introduction

More than half a century ago, autophagy was defined by de Duve to describe the double-membrane vesicle containing cytoplasm and organelles in rat liver cells treated with glucagon [1, 2]. Subsequent studies explored the morphology and functional roles of autophagy. During autophagy process, cytosolic constituents such as damaged organelles and macromolecules are engulfed by a double-membrane autophagosome, and then they fuse with lysosomes for degradation and recycle of the engulfed components within the autolysosome to maintain cellular homeostasis. Up to now, different types of autophagy have been discovered, such as chaperone-mediated autophagy, microautophagy, macroautophagy (hereafter referred to as autophagy), pexopagy, and mitophagy [3, 4]. Although these types have some differences in cellular processes, all of them share an almost identical set of conserved autophagy-related protein (ATG) components, which play vital roles at different steps of autophagic process [5, 6]. The

Hui Gao and Chao Liu contributed equally to this work.

identification of these *ATG* genes bring the key breakthrough in the understanding of the autophagic regulation pathways, such as the mammalian Atg8 homologue, *MAP 1LC3* (also known as *LC3*), and its interacting protein SQSTM1/p62, which are widely used as the markers for autophagic flux measurement [7].

It is well established that spermatogenesis, the transformation of diploid spermatogonial cell into haploid spermatozoa, is modulated by testosterone [8, 9]. In male testis, the Leydig cells, which reside around the seminiferous tubules, are the major source for testosterone production [10]. In mouse, two populations of Leydig cells, the fetal and adult Leydig cells, are identified by their different functions and sources [11]. Fetal Leydig cells are emerged exclusively in clusters that followed the formation of testis cords in E12.5 embryo testis and gradually disappeared in 1 or 2 weeks after birth. Fetal Leydig cells contain large amount of smooth endoplasmic reticulum, tubulo-vesicular mitochondria and small lipid droplets, and acquire typical testosterone synthesis property in 48–72 h after their appearance, which are responsible for the differentiation of the Wolffian duct and external genitalia [12]. Adult Leydig cells originate from the outer layer of mesenchymal fibroblasts around the seminiferous tubules after birth. Unlike fetal Leydig cells, adult Leydig cells (hereafter, referred to as Leydig cells) have four different developmental stages, including stem Leydig cells (postnatal days 1–14, d1–14), progenitor Leydig cells (d14–28), immature Leydig cells (d28–56), and adult Leydig cells (d > d56) [13, 14]. Leydig cells begin to acquire and maintain steroidogenic capacity since their immature stage and produce testosterone for germ cell development and Sertoli cell proliferation [15].

Previous studies have demonstrated that autophagy is extremely active in Leydig cells [16, 17]. However, most of these studies are based on electron microscopy observation and lack of dynamic assessment of autophagic flux, which impede our understanding about the functional role of autophagy in Leydig cells. Thus, a comprehensive measurement of autophagy with different methods in Leydig cells would shed light on our knowledge about the functional role of autophagy in regulating male reproductive physiology. Here, we present a set of protocols for examination and interpretation of autophagy in Leydig cells.

2 Materials

2.1 Immunofluorescence Analysis of Autophagic Components

1. PBS (HyClone, SH30256.01B)
2. 4% PFA: 4% paraformaldehyde (PFA) in PBS, adjust pH to 7.4 with 2 N NaOH
3. Sodium citrate buffer: 1.8 mM citric acid, 8.2 mM sodium citrate, and adjust pH to 6.0 with citric acid

4. 5% BSA: add 5% (w/v) bovine serum albumin (BSA) to PBS and filter after full dissolution
5. Primary antibody solution: 1.5% (w/v) BSA in PBS
6. Secondary antibody solution: diluted aimed secondary antibody in PBS
7. DAPI: stock solution, 5 mg/mL in storage buffer, working at 0.5 mg/mL in PBS in the dark
8. Mounting medium (Solarbio, S2100)
9. 3-Aminopropyl-triethoxysilane (APES) (Zhong Shan Jin Qiao, ZLI-9001)

2.2 Transmission Electron Microscopy

1. 100 mM sodium cacodylate buffer: 5.35 g Na-cacodylic acid in 250 mL distilled water, and adjust pH to 7.3 with 1 N HCl
2. 2.5% Glutaraldehyde: diluted 25% glutaraldehyde (Alfa Aesar, A17876) with distilled water to 2.5%
3. 1.5% PFA: diluted 10% PFA with distilled water to 1.5%, and adjust pH to 7.3 with 2 N NaOH
4. 1% Osmium tetroxide: dissolve 1 g osmium tetroxide (Sigma, 419494) in 100 mL distilled water. Store at 4 °C
5. 2% Uranyl acetate: dissolve 1 mL uranyl acetate (Zhong Jing Ke Yi, GS02625) in 49 mL distilled water
6. Resin: SPI-Pon 812 Epoxy Embedding Kit (SPI Supplies, 02635-AB). Mix 13 mL SPI-Pon 812, 7 mL NMA, 8 mL DDSA, and 10–12 drops DMP-30. Store at 4 °C and keep dry
7. Uranyl acetate for staining: dissolve 1 mL uranyl acetate in 99 mL of distilled water. Store in 10 mL Norm-Ject syringes with a 0.22-μm filter at the tip of the syringe. Avoid the light and keep at 4 °C for months
8. Lead citrate for staining of thin sections: Solution A: 1 M lead nitrate dissolved in distilled water. Solution B: 1 M trisodic sodium citrate dissolved in distilled water. Solution C: 1 N NaOH dissolved in distilled water. Gently mix 1 mL solution A, 1.5 mL solution B, and 8 mL boiled distilled water to form a milk-like liquid; 2 mL of solution C was added and then the mixture turns clear. Store the mixture in 5 mL syringe with a 0.22-μm filter at the tip of the syringe. Store at 4 °C for several months

2.3 Immunoblotting Analysis of LC3 and SQSTM1/p62

1. F12-DMEM: Dulbecco's Modified Eagle's Medium–Ham's Nutrient Mixture F-12 (HyClone, SH30023.01B)
2. Collagenase IV from *Clostridium histolyticum*, type IV (Sigma, C5138)
3. DNase I from bovine pancreas, lyophilized (Sigma, AMPD1)

4. Percoll density gradient buffer: 5 mL 30, 40, 50, and 60% (v/v) Percoll density buffers are made by diluting Percoll (Solarbio, p8370) in HBSS. The wall of Corning High Clarity PP Centrifuge Tube (430790) is rinsed by adding 1 mL fetal calf serum (Gibco, 10270) and then discarded. Add 2 mL 60% Percoll density buffer into the tube, and gently overlay with 2 mL 50% Percoll density buffer, 4 mL 40% Percoll density buffer, and 2 mL 30% Percoll density buffer
5. Trypsin from bovine pancreas (Gibco, 25200072)
6. Leydig cell culture medium: F12-DMEM containing 15% fetal calf serum (Gibco, 10270) and penicillin–streptomycin (Gibco, 15140-122)
7. RIPA lysis buffer: (Solarbio, R0020)
8. Phenylmethylsulfonyl fluoride (PMSF): 1 M stock solution in dimethyl sulfoxide (DMSO) to a final working concentration of 1 mM
9. Protease inhibitor cocktail (Roche Diagnostics, 04693132001)
10. 2× Loading buffer: 100 mM Tris–HCl, pH 6.8, 4% SDS, 20% glycerol, 200 mM dithiothreitol (DTT), and 0.2% bromophenol blue
11. 15% Running gel: 5.0 mL of 30% acrylamide solution was mixed with 2.5 mL of 1.5 M Tris–HCl, pH 8.8, 2.3 mL of distilled water, 0.1 mL of 10% SDS, 100 μL of 10% APS, and 4 μL of TEMED
12. Stacking gel: 0.67 mL of 30% acrylamide solution was mixed with 0.5 mL of 1.0 M Tris–HCl, pH 6.8 [at 25 °C], 2.7 mL of distilled water, 0.04 mL of 10% SDS, 40 μL of 10% APS, and 4 μL of TEMED
13. Running buffer: 25 mM Tris, 192 mM glycine, and 0.1% (w/v) SDS
14. Transfer buffer: 48 mM Tris, 39 mM glycine, and 20% methanol (analytical grade)
15. Blocking buffer: 5% (w/v) Difco Skim Milk (BD, 232100) in PBS
16. Immunoblotting primary antibody solution: 1:1000 anti-LC3 (Sigma, L7543), 1:1000 anti-SQSTM1/p62 polyclonal antibody (CST, 5114) in Primary antibody dilution buffer (Beyotime, P0023A)
17. Immunoblotting secondary antibody solution: diluted secondary antibody 1:10,000 in blocking buffer

2.4 GFP-LC3 Lysosomal Delivery and Proteolysis

1. Opti-MEM: Opti-MEM reduced serum medium (Gibco, 31985-070)
2. Lipofectamine 2000 (Invitrogen, 11668019)

2.5 Turnover of Autophagic Substrate Proteins

1. Cycloheximide (Sigma, R750107)
2. TAP buffer: 10% glycerol, 50 mM HEPES-KOH pH 7.5, 100 mM KCl, 2 mM EDTA, 0.1% NP-40, 10 mM NaF, 0.25 mM Na_3VO_4, 50 mM β-glycerolphosphate, and 2 mM DTT
3. IP buffer: 20 mM Tris, pH 7.4, 2 mM EGTA, and 1% NP-40
4. Protein A-Sepharose (GE, 17-1279-03)

2.6 Antibodies

1. LC3 for immunofluorescence: rabbit anti-LC3B polyclonal antibody (Abcam, ab58610, Cambridge, UK), 1:100
2. LAMP2 for immunofluorescence: rat anti-LAMP2 polyclonal antibody (Abcam, ab13524, Cambridge, UK), 1:100
3. LC3 for immunoprecipitation: mouse anti-LC3B monoclonal antibody (MBL, M152-3 Nagoya, Japan), 1:100
4. Secondary antibodies for immunofluorescence: purchased from Zhong Shan Jin Qiao
5. Secondary antibodies for immunoblotting: Alexa Fluor 680-conjugated goat anti-mouse (Thermo Fisher Scientific, A21057, MA, USA), and Alexa Fluor 680-conjugated goat anti-rabbit (Thermo Fisher Scientific, A21109, MA, USA)

3 Methods

3.1 Immuno-fluorescence Analysis of Autophagic Components

1. Testes are dissected from 6 to 8-week-old mice immediately after euthanasia. Wash testes with PBS immediately, repeat twice. Fix testes in 4% PFA in PBS pH 7.4 at room temperature (RT) for up to 24 h.
2. Embed testes in paraffin.
3. Cut 5-μm thick testis sections and mount on APES pre-coated glass slide.
4. Deparaffinize sections in xylene 3 × 10 min and rehydrate in an alcohol series, thrice in 100% alcohol for 5 min, followed with 95, 80, and 70% each 5 min.
5. Wash briefly for 1 s in PBS and boil for 15 min in sodium citrate buffer for antigen retrieval. Chill the samples at RT.
6. Wash with PBS thrice for 5 min each at RT.
7. Block the samples in 5% BSA for at least 1 h at RT.

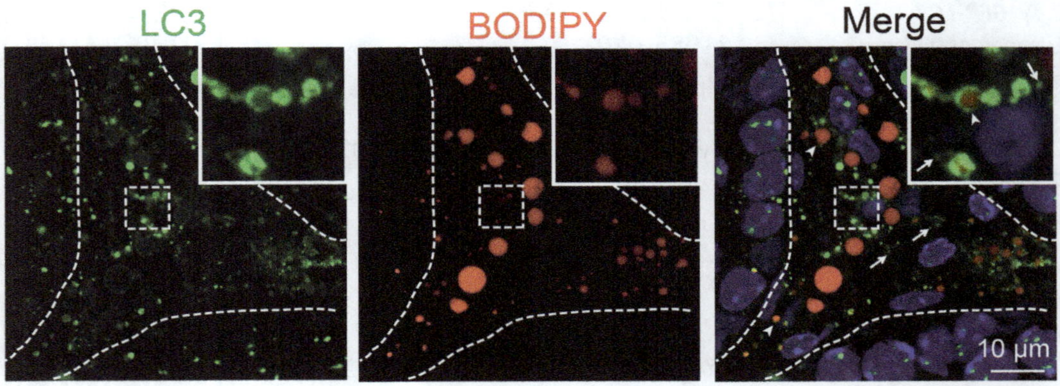

Fig. 1 Immunofluorescence analysis of LC3 and lipid droplets in Leydig cells. LC3 is an important autophagic marker protein, which localizes at different autophagic organelle membrane. In mouse testis, Leydig cells, LC3 (green) was co-localized with lipids (stained with BODIPY, red). DAPI (blue) indicate nuclei. The arrows indicate LC3 puncta out of lipid droplets. Arrowheads show LC3 wrapped lipid droplets

8. Drain the BSA from the glass slide. Dilute the primary antibody with primary antibody solution to desired concentration and add it to the samples. Incubate it at 4 °C overnight (*see* **Note 1**).
9. Discard the primary antibody solution and wash the glass slide thrice with PBS for 5 min each.
10. Drain the PBS from the glass slide. Dilute the secondary antibody in 1:100 to 1:200 with PBS and add it to the cover slip for 1 h at 37 °C.
11. Discard the secondary antibody solution and wash the glass slide thrice with PBS for 5 min each.
12. Drain the PBS from the glass slide and incubate in 1× DAPI solution at RT for 5 min.
13. Wash the cover slips twice with PBS for 5 min each.
14. Drop one drip mounting medium on the samples. Put the cover slips down on the sample and seal the cover slip with nail polish to prevent drying.
15. Image can be viewed by fluorescence microscopy (Fig. 1).

3.2 Transmission Electron Microscopy

The transmission electron microscopy (TEM) is one of the most accurate methods for the observation of autophagy compartments in cells. Contrast to the frequently used light microscopy, the resolution of TEM far surpasses it. However, the small TEM sample size and time-consuming process limit its usage (Fig. 2).

1. Testes were isolated from mouse with surgical scissors after euthanasia. Wash testes with PBS for three times immediately.
2. Gently puncture testes in three to four areas with a 12-guage syringe, taking care not to deform the testes structure, fix the

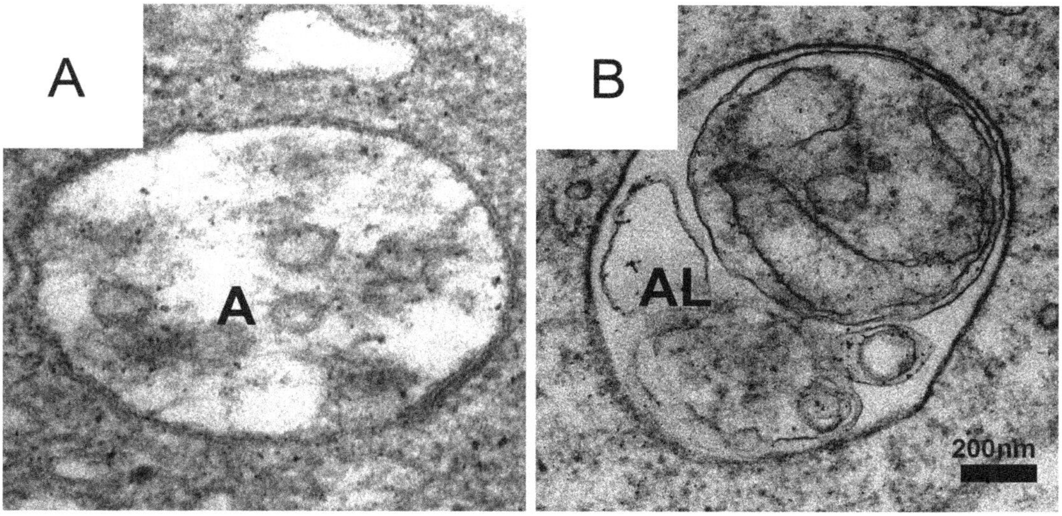

Fig. 2 Transmission electron microscopy (TEM) analysis of autophagosomes and autolysosome in Leydig cells. (**a**) The autophagosome is visible with double membranes. A indicates autophagosome. (**b**) The autolysosome is visible as a single membrane structure and contains degraded organelles. AL indicates autolysosome

testes in 2.5% (v/v) glutaraldehyde, and 1.5% (v/v) PFA in 100 mM sodium cacodylate buffer (pH 7.3) for 4 h at RT.

3. Wash the testis in 100 mM sodium cacodylate buffer (pH 7.3) for 2 × 30 min at RT.
4. Postfix the testis tissue with 1% osmium tetroxide in water for 1 h at RT.
5. Wash testis tissue twice with distilled H_2O, 3 min each.
6. Stain the testis tissue with 2% uranyl acetate in water at room temperature for 1 h in the dark.
7. Drain the water from the sample for resin embedding, and gently mix the testis with a graded acetone series at RT (50, 60, 70, 80, 90, and 95% acetone in water, 5 min each). Incubate with 100% acetone 3 × 5 min in a new tube.
8. To infiltrate with resin, the testis incubates for 2 h with 3:1 100% acetone and resin, followed by 1.5 h in 1:1 100% acetone and resin at RT. Then, incubate for 6 h in 1:3 100% acetone and resin at RT.
9. Incubate for 3 × 12 h in pure resin at RT.
10. Embed the tissue in freshly made resin in a silicone mold. Polymerize the resin at 35 °C for 24 h, 45 °C for 24 h, and then at 65 °C for 24 h.
11. Cut 60 nm sections on an ultramicrotome with a diamond knife, pick the sections on 100-mesh copper grids, and

poststain with 2% uranyl acetate in water for 15 min and 0.3% lead citrate for 3 min.

12. Image can be observed using a JEM-1400 transmission electron microscope.

3.3 Immunoblotting Analysis of LC3 and SQSTM1/p62

LC3 is an important autophagic marker protein, which localizes at different autophagic organelle membrane from the phagophore (precursor stage of autophagosome) to the autolysosome [18]. Once the nascent LC3 (proLC3) was synthesized, it was processed into LC3-I (molecular mass around 16 kDa) immediately [19]. LC3-I could be conjugated to phosphatidylethanolamine (PE) to form LC3-II (approximately 14 kDa), which can be recruited to autophagosomal membranes [20]. When the autophagosome is fused with lysosome to form autolysosome, the LC3-II in the autolysosome inner membrane is degraded by acid hydrolases in the lysosome. Thus, the amount of LC3-II is correlated with the number of autophagosomes, and it is a good marker for measuring autophagic activity [21].

The SQSTM1/p62 protein is another widely used autophagy marker, which is commonly found to bind with LC3 via LC3 interaction region directly. The p62 protein is considered to be an adaptor protein to deliver selective autophagic substrate for degradation and degraded by autophagy in autolysosome. When autophagy is inhibited, p62 quantities increased, which could be served as an indicator of autophagic flux blocking [22].

3.3.1 Primary Leydig Cell Isolation

1. Testes are dissected from 6 to 8-week-old mice immediately after euthanasia. Wash the testes with 75% alcohol for sterilizing, followed by rinses with PBS for three times.
2. Decapsulate the testes, removal of the testes' tunica albuginea with eye scissors carefully, avoid the destruction of seminiferous tubules. Wash the seminiferous tubules with PBS for two times.
3. Incubate the seminiferous tubules with 2 mg/mL collagenase IV and 0.5 mg/mL DNase I in F12/DMEM medium for 15–20 min at 37 °C on a shaker at 85 rpm (*see* **Note 2**).
4. After incubation, equal volume of Leydig cells culture medium was added to halt the digestion of collagenase IV, and the digested medium was sedimented naturally for 2 min at RT and then filtered through a 200-mesh cell strainer.
5. Centrifuge the filtered medium through a swing-out rotor centrifuge at 500 × *g* for 5 min in RT (*see* **Note 3**).
6. After centrifugation, the supernatant was discarded. The pellet was suspended in 2 mL HBSS medium and loaded onto the Percoll density gradient buffer gently.

7. Centrifuge the loaded Percoll density gradient buffer through a swing-out rotor centrifuge at 20,000 × *g* for 1 h at RT, and the primary Leydig cells were enriched at the 50% Percoll fraction.
8. The upper Percoll fraction was removed gently by pipette aspiration and the 50% Percoll fraction was transferred into a 15-mL centrifuge tube, and PBS was added to dilute the Percoll buffer and centrifuge through a swing-out rotor centrifuge at 500 × *g* for 5 min to pellet the Leydig cells (*see* **Note 4**).
9. Discard the supernatant, and suspend the pellet with 15 mL PBS. Centrifuge at 500 × *g* for 5 min to pellet the Leydig cells.
10. The pellet was dissolved with the required volume of Leydig cell culture medium and dispersed into culture dishes, and the Leydig cells were maintained at 34 °C and 5% CO_2.

3.3.2 Preparation of Leydig Cell Lysates

1. Remove the Leydig cell culture medium, and wash the Leydig cells twice with cold (4 °C) PBS. Add trypsin to the culture dish and incubate the cells at 37 °C for 2–3 min. The reaction was terminated by adding equal volume of pre-warmed Leydig cell culture medium. Gently knock the dish to detach the cells. Centrifuge the primary cultured Leydig cells through a swing-out rotor at 4 °C, 500 × g for 5 min.
2. Discard supernatant and suspend the Leydig cell pellets with 80–100 μL of cold RIPA-like buffer supplemented with 1 mM PMSF and 1× protein inhibitor cocktail. Incubate on ice for 30 min, and alternatively, cells can be sonicated briefly with Sonic VCX750 at 25% amplitude for 4 s, and then incubated on ice for 20 min.
3. Centrifuge the homogenates at 4 °C, 13,523 × *g* for 15 min. Transfer the supernatant to a new tube.
4. The protein concentration of whole cell lysates was determined by using the Bio-Rad DC Protein Assay.
5. The protein samples were mixed with equal volume of 2× loading buffer, and boiled for 10 min at 65 °C (*see* **Note 5**).
6. Centrifuge the protein samples at 4 °C, 13,523 × *g* for 2 min and immediately frozen in liquid nitrogen. The samples could be stored at −80 °C.

3.3.3 SDS-PAGE and Immunoblotting

1. The SDS-polyacrylamide running and stacking gels were prepared according to the standard protocol. First, the 15% running gel medium was transferred to the gap between the glass plates with 1.0 mm thickness and then covered with isopropanol to ensure a flat surface without air bubble.
2. The running gel is polymerized at RT for 30 min. Once complete, discard the isopropanol and wash the running gel with water for one time.

3. Prepare the stacking gel mixture and rapidly overlay it onto the polymerized running gel. Carefully insert the comb without air bubble, waiting for the stacking gel to be polymerized.
4. After the polymerization, transfer the glass plates into the SDS-PAGE electrophoresis apparatus; add the Tris-glycine buffer to the apparatus. Remove the comb and rinse the wells with Tris-glycine buffer though syringe to get rid of the gel fragments.
5. Load 10–15 μL samples into the well and run the electrophoresis. After electrophoresis was completed, gently separate the polyacrylamide gel from the glass plates and soak it in the precooled western transfer buffer for 5 min.

3.3.4 Immunoblotting

1. Prepare two sponges, six sheets of filter paper, one nitrocellulose membrane (or PVDF membrane) with the same dimensions of the polyacrylamide gel and soak it in the precooled western transfer buffer.
2. Prepare the transfer sandwich in the following sequence from the bottom to up: one sponge, one sheet of filter paper, the nitrocellulose membrane, the polyacrylamide gel, and one sheet of filter paper, one sponge, then press the transfer sandwich to roll out air bubbles.
3. Connecting the transfer apparatus to power supply, set the transfer apparatus to 25 V, 0.16 A for around 1 h.
4. After the transfer terminated, wash the membrane with PBS. Incubate the membrane in 10 mL of blocking buffer for 30 min at RT on a shaker with 85 rpm.
5. Wash the membrane with PBS for one time and incubate it with diluted immunoblotting primary antibody solution at 4 °C overnight on a shaker.
6. Wash the membrane three times with PBST for 10 min each on a shaker.
7. Incubate the membrane with diluted immunoblotting secondary antibody solution at RT for 1 h on a shaker.
8. Wash the membrane three times with PBST for 10 min each on a shaker.
9. Detect the protein bands with an Odyssey infrared imager (9120, LI-COR Biosciences, Lincoln, NE).

3.4 GFP-LC3 Lysosomal Delivery and Proteolysis

Previous study has confirmed that overexpression of GFP-LC3 shows no influence on the endogenous autophagy activity, which makes it a useful tool to assessing autophagy flux in the cells [23]. For example, real-time observation of GFP-LC3 dots though fluorescence microscope is simple and feasible. In addition, this method also can be applied in immunoblotting. During the autophagy process, the GFP-LC3 fusion protein can be breakdown, whereas the GFP part of the chimera is difficult to be removed. Therefore, the free GFP can be used as an indicator of autophagic activity [24].

1. For GFP-LC3 transfection, primary cultured Leydig cells were seeded in 35 mm^2 cell culture plates at 2×10^6 per plate. After 12–24 h, the cell could reach 60–80% confluency. Change new Leydig cell culture medium 1 h before transfection.
2. Dilute 2.5–5 μg of GFP-LC3 plasmid in 125 μL Opti-MEM, and dilute 7.5–15 μL of Lipofectamine 2000 in 125 μL Opti-MEM in another centrifuge tube. Stew for 5 min and then gently mix the two solutions, set aside for another 15 min (*see* **Note 6**).
3. Add 250 μL transfection mixtures to the cell for 12 h at 34 °C, 5% CO_2.
4. Discard the medium and add 2 m L Leydig cell culture medium for 36 h.
5. Collect primary cultured Leydig cells by trypsinization and perform the immunoblotting analysis of GFP as described in Sects. 3.3.2–3.3.4.

3.5 Turnover of Autophagic Substrate Proteins

3.5.1 Immunoblotting Analysis of Autophagic Substrate Proteins

1. Isolate the primary Leydig cells from mouse testes as described in Sect. 3.3.1.
2. Prepare Leydig cell lysates as described in Sect. 3.3.2.
3. Perform the immunoblotting analysis of autophagic substrate proteins as described in Sects. 3.3.3 and 3.3.4.

3.5.2 Immunofluorescence Analysis of Autophagic Substrate Proteins

1. Isolate the primary Leydig cells from mouse testes as described in Sect. 3.3.1. Seed Leydig cells to 35 mm^2 culture dishes in which sterile cover slips have been previously deposited.
2. Treat Leydig cells for indicated times. Discard the growth medium and wash the cells with pre-warmed PBS.
3. Carefully take out the cover slips from PBS with a forceps, and put it on a glass slide with cell side up.
4. Fix Leydig cells in 4% PFA for 5 min at RT.
5. Remove fixative solution and wash the cover slips thrice with PBS for 5 min each.
6. Allow Leydig cells to be blocked in 5% BSA for at least 1 h in RT.

7. Drain the BSA from the cover slips. Dilute the primary antibody with primary antibody solution to desired concentration and add it to the cover slips. Incubate it at 4 °C overnight.
8. Discard the primary antibody solution and wash the cover slips thrice with PBS for 5 min each.
9. Drain the PBS from the cover slips. Dilute the secondary antibody in 1:100 to 1:200 with PBS and add it to the cover slip for 1 h 4 °C at 37 °C.
10. Discard the secondary antibody solution and wash the cover slips thrice with PBS for 5 min each.
11. Drain the PBS from the cover slips and incubate in 1× DAPI solution at room temperature for 5 min.
12. Wash the cover slips twice with PBS for 5 min each.
13. Mount the cover slips with Leydig cells down onto glass slide using mounting medium and seal the cover slip with nail polish to prevent drying.
14. Image can be viewed by fluorescence microscopy.

3.6 Cycloheximide Chase Assay for Autophagic Substrate Protein

The cycloheximide chase (CHX) is frequently used in protein turnover assays as a protein synthesis inhibitor [25]. When cells were treated with CHX to eliminate newly synthesized proteins, the immunoblotting could be used to monitor preexisting protein degradation speed [26].

1. Isolate the primary Leydig cells from mouse testes as described in Sect. 3.3.1. Seed 2 × 10^6 Leydig cells to 35 mm^2 culture dish 24 h before experiment.
2. When the cell grows to 60–80% confluency, discard the old culture medium. Wash the cells with PBS for one time. Add 100 μg/mL CHX to the culture medium and treat the cells for different time points.
3. Collect the cultured Leydig cells lysates as described in Sect. 3.3.2.
4. Perform the immunoblotting analysis of the targeted protein as described in Sects. 3.3.3 and 3.3.4.

3.7 Co-immunoprecipitation of LC3 and Autophagic Substrate Proteins

1. Removed the culture medium, and wash the Leydig cells twice with cold (4 °C) PBS. Collect primary cultured Leydig cells by trypsinization and centrifuge the cell through a swing-out rotor centrifuge at 4 °C, 500 × *g* for 5 min.
2. Discard the supernatant and suspend the pellets by using 150–200 μL of TAP buffer supplemented with 1 mM PMSF and a protein inhibitor cocktail. Incubate the samples on ice for 30 min.

3. Centrifuge the homogenates at 4 °C, 13,523 × *g* for 15 min. Transfer the supernatant to a new tube, take 10% supernatant as input samples, freeze them with liquid nitrogen immediately, and store at −80 °C.
4. Incubate the remnant cell lysates with primary antibody at 4 °C overnight.
5. Wash protein A-Sepharose beads with ddH_2O at 4 °C, 3000 rpm for 2 min, use TAP buffer to wash the beads at 4 °C, 3000 rpm for 2 min, repeat two times. Add beads into the cell lysates and incubate them for 2 h at 4 °C on a shaker at 60 rpm.
6. Centrifuge the mixture at 4 °C, 3000 rpm for 2 min. Remove the supernatant and wash the collected beads with 1 mL TAP buffer, incubate them for 5 min at 4 °C on a shaker at 60 rpm, and centrifuge at 4 °C, 3000 rpm for 2 min to pellet the beads (*see* **Note 7**).
7. Repeat the wash step for two times.
8. After the final wash, discard the supernatant, add 30 μL of 2× loading buffer to beads and the input samples, and incubate for 15 min at 65 °C.
9. Chill the samples on ice and centrifuge at 4 °C, 13,523 × *g* for 10 min.
10. Perform the immunoblotting analysis of the targeted protein as described in Sects. 3.3.3 and 3.3.4.

4 Notes

1. Each primary antibody requires different working concentration, so it is very important to test the sensitivity and accuracy of the primary antibody in preliminary experiment. In addition, as many autophagic marker proteins are membrane-associated proteins, the addition of detergent in the antibody dilution buffer needs to be carefully considered.
2. The digestion of seminiferous tubules is the key step in primary Leydig cell isolation. We suggest to observing the digestion process continuously, once the seminiferous tubules showed filamentous dispersion, it should be terminated quickly to avoid excessive digestion which would induce the release of Sertoli cells and germ cells from the seminiferous tubules.
3. During this process, a dispersed line of the Leydig cells but not pellet could be observed after centrifugation by using some angle rotors. To avoid this, we suggest the researchers to use a swing-out rotor to collect the cell.

4. Gently aspirate the 50% Percoll fraction from the Percoll gradient tubes. Try to avoid aspirating other fractions.
5. Many Atg protein and adaptor proteins are membrane-associated proteins, and we recommend denaturing the proteins at 65 °C rather than 95 °C.
6. As the transfection rate of Leydig cells is lower, the concentration of plasmid and Lipofectamine 2000 should be excessive.
7. As the binding of LC3 and other autophagic substrate proteins is complicated. The incubation time and wash times should be based on the individual circumstance.

Acknowledgements

This work was supported by the National Nature Science of China (Grant No. 91649202 and 31471277) and National key R&D program of China (Grant No. 2016YFA0500901). Hui Gao and Chao Liu contributed equally to this work.

References

1. Yang Z, Klionsky DJ (2010) Eaten alive: a history of macroautophagy. Nat Cell Biol 12(9):814–822
2. De Duve C, Wattiaux R (1966) Functions of lysosomes. Annu Rev Physiol 28:435–492
3. Kim I, Rodriguez-Enriquez S, Lemasters JJ (2007) Selective degradation of mitochondria by mitophagy. Arch Biochem Biophys 462(2):245–253
4. Singh R et al (2009) Autophagy regulates lipid metabolism. Nature 458(7242):1131–1135
5. Ravikumar B et al (2010) Regulation of mammalian autophagy in physiology and pathophysiology. Physiol Rev 90(4):1383–1435
6. Mizushima N, Yoshimori T, Ohsumi Y (2011) The role of Atg proteins in autophagosome formation. Annu Rev Cell Dev Biol 27:107–132
7. Jiang P, Mizushima N (2015) LC3- and p62-based biochemical methods for the analysis of autophagy progression in mammalian cells. Methods 75:13–18
8. Skinner MK (1991) Cell-cell interactions in the testis. Endocr Rev 12(1):45–77
9. Sharpe RM (1987) Testosterone and spermatogenesis. J Endocrinol 113(1):1–2
10. Haider SG (2004) Cell biology of Leydig cells in the testis. Int Rev Cytol 233:181–241
11. Habert R, Lejeune H, Saez JM (2001) Origin, differentiation and regulation of fetal and adult Leydig cells. Mol Cell Endocrinol 179(1–2):47–74
12. Kuopio T, Paranko J, Pelliniemi LJ (1989) Basement membrane and epithelial features of fetal-type Leydig cells in rat and human testis. Differentiation 40(3):198–206
13. Risbridger GP, de Kretser DM (1986) Percoll-gradient separation of Leydig cells from postnatal rat testes. J Reprod Fertil 76(1):331–338
14. Mendis-Handagama SM, Ariyaratne HB (2001) Differentiation of the adult Leydig cell population in the postnatal testis. Biol Reprod 65(3):660–671
15. Sharma RS, Pal PC, Rajalakshmi M (2006) Isolation and culture of Leydig cells from adult rats. Indian J Clin Biochem 21(1):27–33
16. Tang XM, Clermont Y, Hermo L (1988) Origin and fate of autophagosomes in Leydig cells of normal adult rats. J Androl 9(4):284–293
17. Yi J, Tang XM (1995) Functional implication of autophagy in steroid-secreting cells of the rat. Anat Rec 242(2):137–146
18. Kuma A, Matsui M, Mizushima N (2007) LC3, an autophagosome marker, can be incorporated into protein aggregates independent of autophagy: caution in the interpretation of LC3 localization. Autophagy 3(4):323–328
19. Kabeya Y et al (2004) LC3, GABARAP and GATE16 localize to autophagosomal

membrane depending on form-II formation. J Cell Sci 117(Pt 13):2805–2812
20. Kabeya Y et al (2000) LC3, a mammalian homologue of yeast Apg8p, is localized in autophagosome membranes after processing. EMBO J 19(21):5720–5728
21. Mizushima N, Yoshimori T (2007) How to interpret LC3 immunoblotting. Autophagy 3 (6):542–545
22. Mizushima N, Yoshimori T, Levine B (2010) Methods in mammalian autophagy research. Cell 140(3):313–326
23. Mizushima N (2004) Methods for monitoring autophagy. Int J Biochem Cell Biol 36 (12):2491–2502
24. Mizushima N (2009) Methods for monitoring autophagy using GFP-LC3 transgenic mice. Methods Enzymol 452:13–23
25. Zhang L et al (2007) Method for real-time monitoring of protein degradation at the single cell level. BioTechniques 42(4):446, 448, 450
26. Yewdell JW, Lacsina JR, Rechsteiner MC, Nicchitta CV (2011) Out with the old, in with the new? Comparing methods for measuring protein degradation. Cell Biol Int 35(5):457–462

Methods in Molecular Biology (2018) 1854: 87–103
DOI 10.1007/7651_2017_84

Published online: 4 November 2017

Simultaneous Detection of Autophagy and Epithelial to Mesenchymal Transition in the Non-small Cell Lung Cancer Cells

Javad Alizadeh, Shahla Shojaei, Adel Sepanjnia, Mohammad Hashemi, Eftekhar Eftekharpour, and Saeid Ghavami

Abstract

Autophagy is increasingly identified as a central player in many cellular activities from cell proliferation to cell division, migration, and differentiation. However, it is also considered as a double-edged sword in cancer biology which either promotes oncogenesis/invasion or sensitizes the tumor cells to chemotherapy induced apoptosis. Recent investigations have provided direct evidence for regulation of cellular phenotype via autophagy pathway. One of the most important types of phenotype conversion is Epithelial-Mesenchymal-Transition (EMT), resulting in alteration of epithelial cell properties to a more mesenchymal form. In the current chapter, we provide a method which is established and being used in our laboratory for detection of autophagy and EMT in lung epithelial cells and show the involvement of autophagy in modulation of cellular phenotype.

Keywords Epithelial to mesenchymal transition, Phenotype conversion, Autophagy, Lung cancer

1 Introduction

Autophagy is a highly regulated and evolutionary conserved process and is closely depended on lysosomes [1]. Autophagy has an important role in maintaining the cellular homeostasis at an optimal state by sensing the source of stress and responding to it by degrading deleterious agents [2, 3]. Likewise, autophagy determines the outcome of a variety of diseases, such as cancer and neurodegeneration, by making modifications in intracellular catabolic activity and recruiting several cellular systems in different tissues [4–8].

Macroautophagy, Microautophagy, and Chaperone-Mediated Autophagy (CMA) are three major types of autophagy (Fig. 1) [9]. A double membrane autophagosome engulfs the cargo, a damaged organelle or long-lived cytoplasmic proteins, and fuses to lysosome subsequently to degrade the cargo through Macroautophagy [10]. Through the Microautophagy, the substrate is directly engulfed by invagination of lysosomal or endosomal membrane and then is degraded by lysosomal proteases [11]. The third

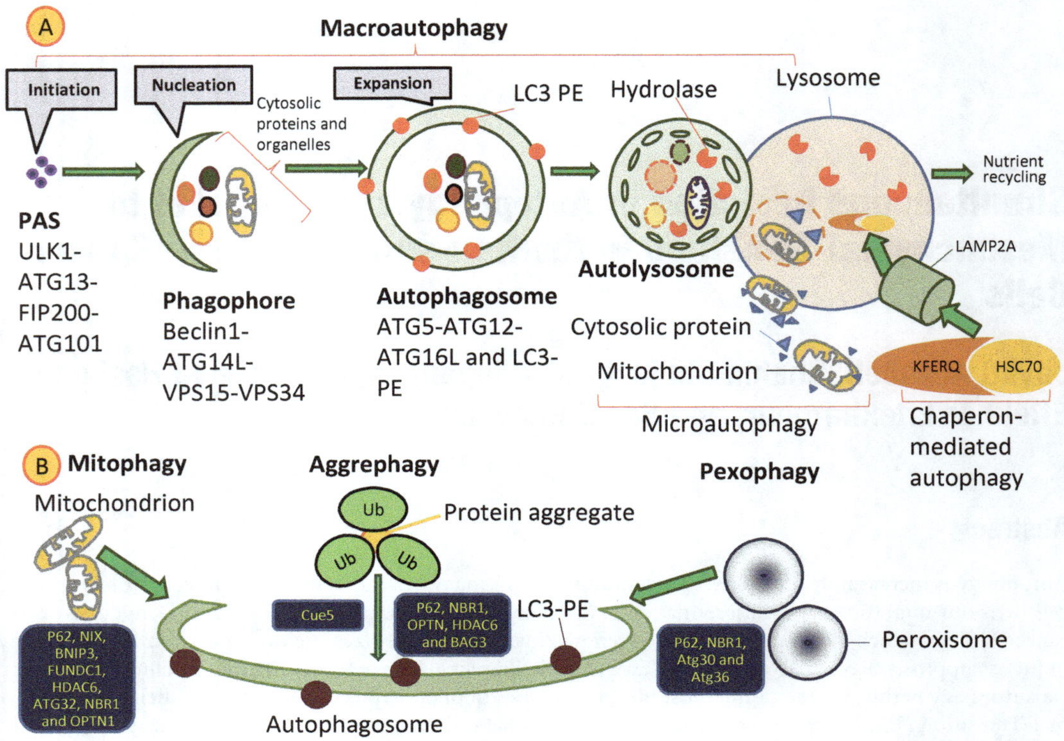

Fig. 1 (**a**) Autophagy is initiated by the nucleation of an isolation membrane or phagophore. This membrane then elongates and closes on itself to form an autophagosome. Growth factors, such as insulin, bind to membrane receptors to activate class I PI3K. This process generates PI (3–5) P3, which recruits protein kinase B (PKB/Akt) and its activator PDK1 (phosphoinositide-dependent kinase 1) to the plasma membrane, resulting in activation of PKB/Akt. Active PKB/Akt indirectly activates mTOR through inhibition of negative regulators [tuberous sclerosis complex (TSC1/2)] of mTOR and activating the mTOR activator Rheb (Ras homologue enriched in brain). The Beclin 1 complex contributes to the nucleation of the phagophore. Beclin 1 complex is regulated by Bcl-2. Elongation of the phagophore membrane is dependent on the Atg12 and LC3 conjugation systems. Closure of the autophagosome is dependent on the activity of the LC3-conjugation system. The autophagosome matures by fusing with endosomes and lysosomes, finally forming the autolysosome where the cargo degradation occurs. (**b**) Specific organelle or protein phagy is involved in removal of damaged proteins and organelle in the mammalian cells

type of autophagy namely CMA functions in a highly selective way without using any membrane to engulf cargos. Proteins targeted by CMA contain a pentapeptide motif including KFERQ (Lys-Phe-Glu-Arg-Gln) sequence that is detected by cytosolic heat shock cognate 70 kDa protein (hsc70). The hsc70 is involved in cargo delivery into the lysosomes through interaction with lysosomal-associated membrane protein type 2A (LAMP2A) receptor [12].

Autophagy plays an essential function in the regulation of cellular phenotype including epithelial and mesenchymal phenotypes of the cell [13] and involved in regulation of epithelial mesenchymal transition (EMT) in many models (Fig. 2) [14, 15].

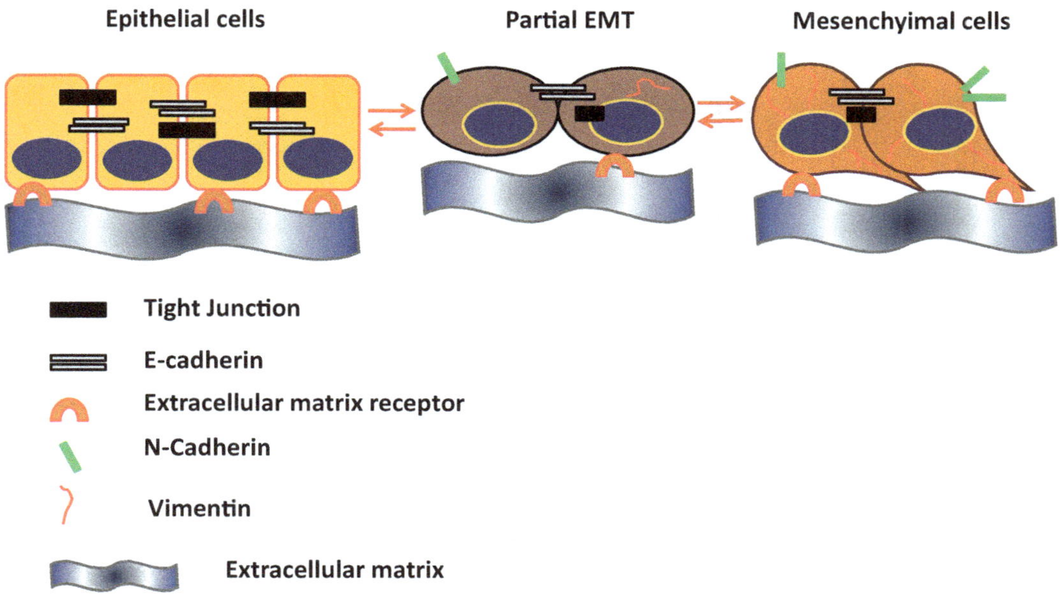

Fig. 2 EMT induces cellular phenotype changes. The most important characterization of epithelial cells is the presence of apico-basal polarity, intact cell–cell contacts, barrier function, and lack of movement. During the EMT process cell–cell contacts are being leaky with appearance of the incipient presence of a front-rear polarity. At the same time upregulation of the mesenchymal markers vimentin and N-cadherin begin. In addition, these partial EMT cells express lower amounts of E-cadherin than epithelial cells

Transforming growth factor beta one ($TGF\beta_1$) superfamily is a well-studied EMT inducer which has three different ligands including $TGF\beta_{1-3}$ [16] that convey their signal through their relevant receptors (TGFβ.R). Using various adaptor and cofactor proteins, $TGF\beta_1$ signaling pathway acts in a wide spectrum of context-dependent functions in different tissues [17] among which invasion and metastasis of cancer cells are important in cancer studies [18, 19]. Figure 3 depicts a number of signaling pathways that are involved in EMT induction [20].

$TGF\beta_1$ interacts with other signaling pathways such as sonic Hedgehog [21], Ras-MAPK [22], Notch [23], and Wnt [24] to induce EMT. Schematic interaction of $TGF\beta_1$ with other signaling systems is shown in Fig. 4 [25]. The pivotal role of $TGF\beta_1$ is well documented in progression and aggressiveness of lung tumors especially by induction of EMT through regulating different proteins including Snail [26], E-cadherin [27], vimentin [28–30], and N-cadherin [30]. $TGF\beta_1$ can also induce EMT in lung cancer by regulating the expression level of integrin [26, 31, 32]. As metastasis is the main cause of death in lung cancer patients [33], an in-depth understanding of EMT regulation seems to be of significant clinical importance.

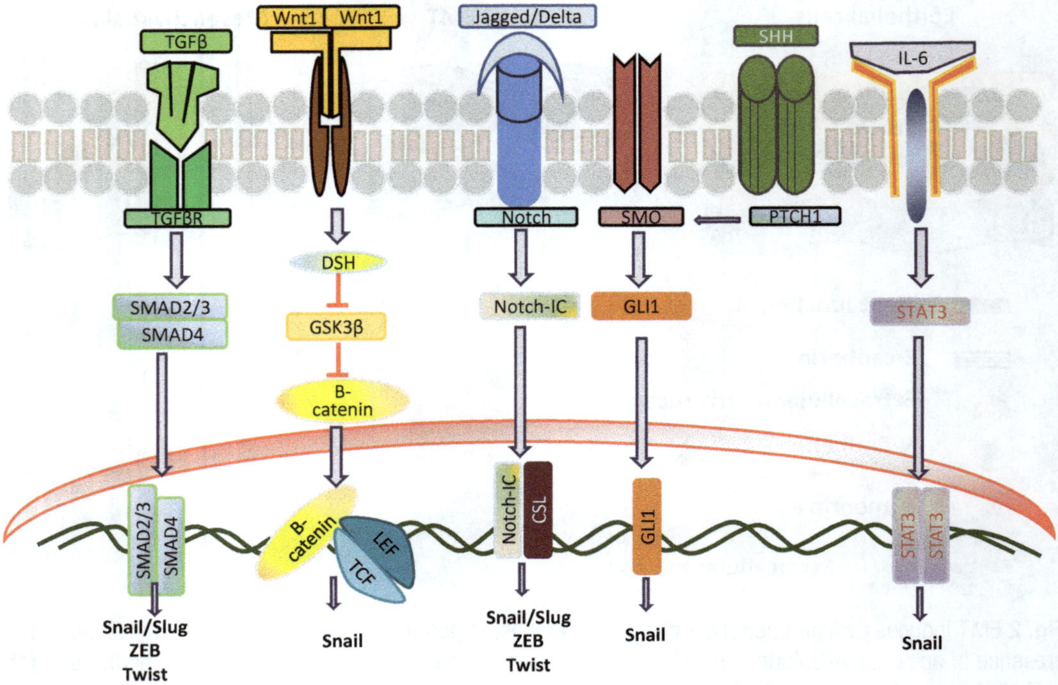

Fig. 3 Signaling pathways and their activation result in EMT process [20]. Diverse signaling pathways associated with epithelial–mesenchymal transition (EMT). Transforming growth factor beta (TGFβ) signals activate SMAD2 and SMAD3 that complex with SMAD4. The trimeric SMAD complex enters the nucleus and leads to the transcription of EMT transcription factors (EMT-TFs). Activation of Wnt signaling inhibits the destruction complex containing glycogen synthase kinase 3 beta (GSK-3β) through Disheveled (DSH), facilitating-catenin to enter the nucleus and activate the Snail transcription, a transcription factor that drives cell toward EMT. Notch receptors can be activated by binding to Delta and Jagged ligands. After activation, Notch intracellular domain (Notch-IC) is released through a cascade of proteolytic cleavages and activates CSL transcription factor to express EMT-TFs. In Sonic Hedgehog (SHH) signaling, ligand binding to Patched 1 (PTCH1) receptors activates Smoothened (SMO) and Glioma (GLI) family transcription factors that induce Snail expression. Interleukin-6 (IL-6) can induce Snail expression by activating STAT3

Autophagy can play both pro- and anti-oncogenic roles in cancer development particularly in the invasion and motility of metastatic cancer cells [34–36]. Accordingly, induction or inhibition of autophagy can be harnessed as a therapeutic intervention against cancer, although this might be affected by cancer type, microenvironment, and cancer stage. In this context, autophagic degradation of EMT promoter protein Snail can drive cells toward MET and consequently have a protective role in some cancers [37]. Using animal models with genetically manipulated autophagy has indicated that in cancer cells tolerating hypoxic conditions induces autophagy as an alternative supplier of energy that is another beneficial role of autophagy for cancer cells beside its role in EMT [38–44]. EMT and autophagy are interconnected [45–51], for example autophagy is involved in hepatocellular

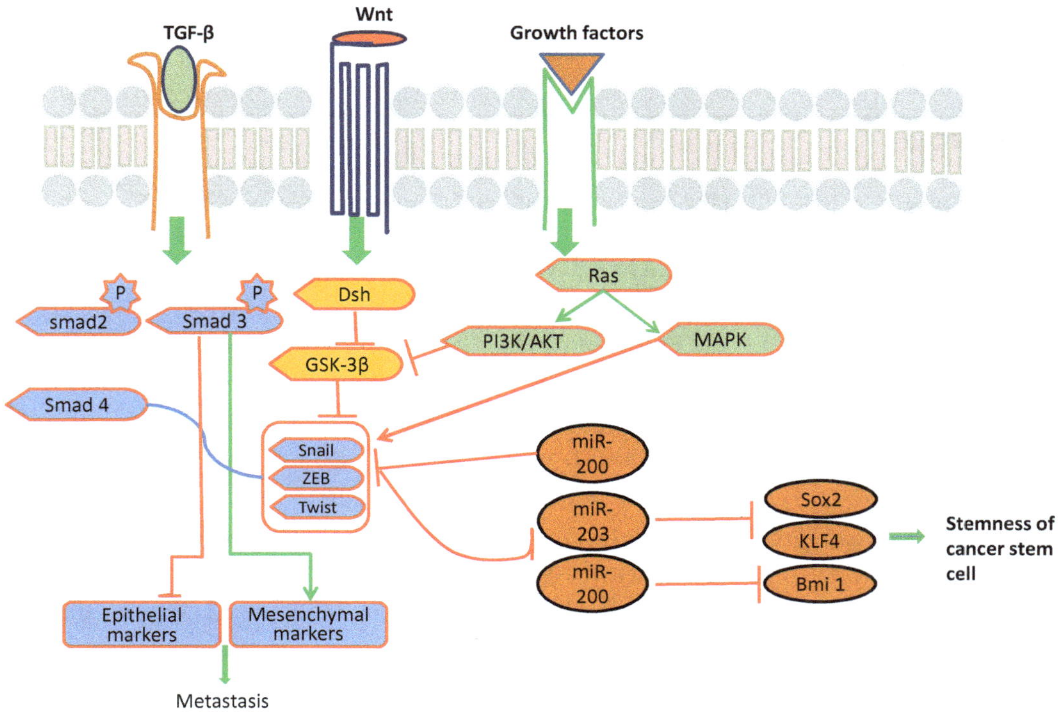

Fig. 4 TGFß signaling, regulation, and association with other pathways which can influence the EMT process [25]. TGFß is a major inducer of EMT. It binds to the receptors leading to the phosphorylation of SMAD2 and SMAD3. The activated forms of SMAD2 and SMAD3 make trimers with SMAD4 and then translocated into nucleus where they associate and cooperate with DNA binding transcriptional factors such as Snail, ZEB, and Twist to regulate the expression of TGFß target genes, resulting in the downregulation of epithelial markers and the upregulation of mesenchymal markers. TGFß also cooperates with other signal factors such as Wnt and growth factors that act through receptor tyrosine kinase to regulate EMT. Several microRNAs have been identified to regulate EMT. miR-200 suppresses EMT mainly through targeting ZEB factors and ZEB factors also regulate the expression of miR-200 and miR-203, linking the EMT and stem maintenance of cancer stem cells

carcinoma invasiveness and metastasis by activating TGFβ1 signaling pathway and subsequent EMT induction [50]. Elevated levels of LC3B-II punctate, which has direct relevance to autophagy, are associated with high metastasis and low survival rate in breast cancer patients [42, 43, 52, 53]. ULK2 is an inducer of autophagy (Beclin1 phosphorylation) which downregulates E-cadherin and induces EMT, resulting in a more invasive phenotype and metastasis [54]. Likewise, another report has shown that higher level of autophagyic factors in glioblastoma patients was associated with more aggressive tumors belonging to the mesenchymal subclass [55]. Contribution of autophagy and EMT to drive cancer stem cells (CSCs) phenotype and their role in the metastasis and invasiveness of CSCs [56] is a new and charming explanation for capacity of CSCs to disseminate heterogeneous tumors to the secondary organs that needs to be examined more extensively [41].

In the following sections we describe the routine methods to investigate autophagy and EMT in lung epithelial cells.

2 Materials

2.1 Cell Lines

1. Human non-small cell lung carcinoma (NSCLC) cell lines A549 (ATCC Number: CCL-185) and H1975 (ATCC Number: CRL-5908).
2. Dulbecco's Modified Eagle's Medium (DMEM) (CORNING; Cat #: 50-003-PB).
3. 10% Fetal Bovine Serum (FBS) (Gibco™; Cat #: 16000044).
4. Insulin/Transferrin/Selenium (ITS) (1%) (Gibco™; Cat #:41400045).
5. Polybrene (Santa Cruz; sc-134220).
6. Tetrazolium dye MTT 3-(4,5-dimethylthiazol-2-yl)-2,5-diphenyltetrazolium bromide (SIGMA; #M2128).
7. 5 ng/ml Recombinant Human TGFβ_1 (rhTGFβ_1)(Catalogue #240-B).

2.2 Western Blotting

1. NP-40 lysis buffer [20 mM Tris–HCl (pH 7.5), 0.5% Nonidet P-40, 0.5 mM PMSF, 100 μM β-glycerol 3-phosphate, and 0.5% protease inhibitor cocktail].
2. 10–15% polyacrylamide gels.
3. Immuno-Blot PVDF Membranes (Bio-Rad; #1620177).
4. Transfer buffer (500 nM glycine, 50 mM Tris–HCl, and 20% methanol).
5. 5% Non-fat dried milk.
6. 1× Tris-buffered saline containing Tween (TBS/0.01% tween 20; TBST).
7. Primary antibodies.
   (a) Anti-LC3B-I antibody 18 kDa 1:1500 dilution SIGMA; # L7543.
   (b) Anti-LC3B-II antibody 16 kDa 1:1500 dilution SIGMA; # L7543.
   (c) Anti-ATG7 antibody 7 kDa 1:1000 Cell signaling; #8558.
   (d) Anti-P62 antibody 62 kDa 1:1000 Cell signaling; #8025.
   (e) Anti-Vimentine antibody 57 kDa 1:1000 Cell signaling; #5741.
   (f) Anti-E-cadherin antibody 135 kDa 1:1000 Cell signaling; #3195.
   (g) Anti-Snail antibody 29 kDa 1:1000 Cell signaling; #3879.

(h) Anti-slug antibody 30 kDa 1:1000 Cell signaling; #9585.
(i) Anti-GAPDH antibody 37 kDa 1:2500 Santa Cruz; sc-47724.

8. Horseradish peroxidase (HRP).
9. Secondary antibodies.
   (a) Anti-Rabbit IgG (whole molecule)–Peroxidase antibody produced in goat 1:5000 dilution SIGMA; # A6154.
   (b) Anti-Mouse IgG (Fab specific)–Peroxidase antibody produced in goat 1:3000 dilution SIGMA; # A-8924.
10. Chemiluminescence (ECL) regents (Amersham-Pharmacia Biotech).

2.3 Fluorescence Live Cell Imaging

1. LC3-GFP plasmid (Addgene; #24920).
2. Effectene Transfection Reagent (QIAGEN; #301425).
3. ITS (1%).
4. Lysotracker Red (Molecular Probes™; LysoTracker® Red DND-99; L7528).
5. DAPI (4′, 6-diamidino-2-phenylindole) (ProLong® Gold Antifade Mountant with DAPI; Cat #: P36931).
6. 1× PBS.

2.4 Immunocytochemistry (ICC)

1. Coated slides (Fisherbrand™ Superfrost™ Plus Microscope Slides; Cat#: 12-550-15).
2. Baf-A1 (0.1 nM) or Rapamycin (500 μM).
3. TGFβ_1 (5 ng/ml).
4. 3.7% formaldehyde.
5. 0.25% triton-X100.
6. Rabbit anti-Vimentin IgG primary antibody (Cell signaling; #5741).
7. Secondary antibodies.
8. Antifade Mountant.

3 Method

3.1 Cell Culture

1. Human non-small cell lung carcinoma (NSCLC) cell lines A549 (ATCC Number: CCL-185) and H1975 (ATCC Number: CRL-5908) were cultured in Dulbecco's Modified Eagle's Medium (DMEM) (CORNING; Cat #: 50-003-PB) with 10% Fetal Bovine Serum (FBS) (Gibco™; Cat #: 16000044). Insulin/Transferrin/Selenium (ITS) (1%) (Gibco™; Cat #:41400045) was used to starve the cells ***(note 1)***.

2. Cells were maintained in a humidified incubator with 95% air and 5% CO_2 at 37 °C (standard cell culture incubator conditions).
3. Tetrazolium dye MTT 3-(4, 5-dimethylthiazol-2-yl)-2, 5-diphenyltetrazolium bromide (SIGMA; #M2128) was used for the cytotoxicity assay ***(note 2)***.

3.2 Production of Stable ATG7 Knock Down (KD) NSCLC Cell Lines

1. Human NSCLC cell lines A549 and H1975 were seeded at a density of 5 × 10^4 cells/well in 12-well plates and cultured in DMEM, 10% FBS for 24 h.
2. After reaching 40% confluency, cells were treated with 10 μg/ml polybrene (Santa Cruz; sc-134220) in DMEM medium (without FBS and) for 1 h ***(note 3)***.
3. Cells, then, transfected with shRNA Lentiviral Particle for *ATG7* and scrambled control (a negative control strategy that has the same nucleotide composition, but not the same sequence, as the test shRNA) both carrying the coding gene for puromycin resistance (Santa Cruz; sc-41447-V, *APG7* shRNA (h) Lentiviral Particles). Cells were transfected at 3 and 6 multiplicity of infections (MOI) for 12 h, followed by replenishing the medium for recovery for 24 h ***(note 4)***.
4. After the recovery, cells that incorporated the shRNA plasmid were selected using Puromycin dihydrochloride (4 μg/ml) (Santa Cruz; sc-108071) containing medium. Cells with the activated puromycin resistance gene survived the selection and clones were isolated ***(note 5)***.
5. The *ATG7* status was later checked by Western blotting in shRNA transfected and scramble cells based on published original research papers [57–60].

3.3 Western Blotting

Western blot analysis was used to detect markers of autophagy and EMT and Smad phosphorylation in A549, H1975, and *ATG7* KD cells. We examined markers of Smad signaling, autophagy, and EMT while GAPDH was used to normalize the results. We followed the procedures used in previous studies for protein assay, sample preparation, and SDS-PAGE [60–62].

1. Briefly, cells were washed and protein extracts were prepared in NP-40 lysis buffer [20 mM Tris–HCl (pH 7.5), 0.5% Nonidet P-40, 0.5 mM PMSF, 100 μM β-glycerol 3-phosphate and 0.5% protease inhibitor cocktail] and samples were stored at −20 °C until used for western blotting.
2. After centrifugation (13,000 × *g* for 10 min) supernatant protein content was determined by Lowry protein assay.
3. Proteins were size fractionated by SDS-PAGE; samples were heated at 90 °C for 5 min and 15–20 μl of samples was loaded

on a 10–15% polyacrylamide gels based on the molecular weight of the proteins, 10 μl of precision plus protein were used as marker standard (Thermo Fischer Scientific, ON, Canada).

4. After electrophoresis, separated proteins were transferred onto Immuno-Blot PVDF Membranes (Bio-Rad; #1620177) under reducing conditions in transfer buffer (500 nM glycine, 50 mM Tris–HCl, and 20% methanol) for 2 h, RT at 100 V.
5. Membranes were blocked with 5% non-fat dried milk and 1× Tris-buffered saline containing Tween (TBS/0.01% tween 20; TBST) at cold room overnight or RT for 1 h.
6. After blocking, membranes were incubated at 4 °C overnight with appropriate primary antibodies in 1% milk in 1× TBST.
7. After overnight incubation membranes were washed three times with 1× TBST for 20 min and incubated with appropriate horseradish peroxidase (HRP) coupled secondary antibodies for 2 h at RT.
8. Membranes were washed again three times for 20 min and incubated with enhanced chemiluminescence (ECL) reagents (Amersham-Pharmacia Biotech) for 2–3 min.
9. The signals were visualized by autoradiography. Obtained protein bands were evaluated for changes in the autophagy, EMT, and Smad signaling pathway and to draw a conclusion and comparison on the role of autophagy inhibition and induction in EMT process in A549 and H1975 cells.
10. The blots for both cells were quantified using dosimetry software Alpha Ease FC.

3.4 Fluorescence Live Cell Imaging Using LC3-GFP

Previous works have used LC3-GFP in order to detect the autophagosomes [63]. This method was used to confirm the autophagy and EMT in presence and absence of TGFβ_1. It involves the transfection of A549 and H1975 cells with LC3-GFP plasmid (Addgene; #24920) using the Effectene Transfection Reagent (QIAGEN; #301425) ***(note 6)***. After transfection, the LC3-GFP plasmid expresses the mammalian LC3 fused to EGFP in the cells which can be detected as green under the immunofluorescent microscopy.

1. A number of 30,000 cells/ml were counted and seeded on 6-well plates (2 ml of medium and cells) and grown to 50% confluency by incubation for 24 h.
2. The transfection of the cells was done with Effectene Transfection Reagent, according to the manufacturer's protocol.
3. After 18 h incubation transfected cells were checked under a fluorescent microscope (Olympus, Markham, ON, Canada) and after confirmation of the transfection their medium was replenished with ITS (1%) for 24 h and treated with TGFβ_1 (5 ng/ml).

4. In the 36 h time point, lysosomes and DNA were stained for 30 minutes with Lysotracker Red (Molecular Probest™; Lyso-Tracker® Red DND-99; L7528) at a concentration of 50 nM and a drop of DAPI (4', 6-diamidino-2-phenylindole) (Pro-Long® Gold Antifade Mountant with DAPI; Cat #: P36931), respectively. DAPI stains the nucleus and also protects fluorescent dyes from fading.
5. After washing the cells with 1× PBS, images were acquired using the fluorescent microscope. LC3 punctate localized with activated lysotracker red identified as autophagy cells. These results validated the WB analysis results.

3.5 Immunocytochemistry (ICC)

To detect the vimentin changes in A549 and H1975 cells and also their corresponding *ATG7* knock down cells we used a combination of LC3-GFP and regular immunocytochemistry protocol according to previous studies [58, 64]. After cells transfected with the LC3-GFP, they were cultured on coated slides (Fisherbrand™ Superfrost™ Plus Microscope Slides; Cat#: 12-550-15) (approximately 5000 cells/spot) in DMEM media (high glucose) with 10% FBS media conditions (using standard cell culture incubator conditions). Cells were starved with ITS (1%) for 24 h and pre-treated with indicated concentrations of Baf-A1 (0.1 nM) or Rapamycin (500 μM). Then, cells co-treated with $TGF\beta_1$ (5 ng/ml) for 36 h. In the indicated time point, immunocytochemistry was done with the following protocol.

1. Cells were fixed with 3.7% formaldehyde for 20 min at RT and washed 3 times for 10 min with 1× TBS.
2. Then, cells were permeabilized with 0.25% triton-X100 for 15 min and blocked with 5% Normal Donkey Serum (NDS) (JACKSON; #017-000-121) for 1 h at RT.
3. Cells were then incubated overnight in 4 °C with the rabbit anti-Vimentin IgG primary antibody (Cell signaling; #5741) (1:200) and corresponding isotype control immunoglobulin (IgG's) diluted in 5% NDS to detect the changes in the EMT marker vimentin.
4. Slides were washed three times for 10 min and incubated with appropriate fluorochrome-conjugated secondary antibodies (Alexa Fluor conjugated) for 1 h at RT and washed three times for 10 min.
5. Slides were then incubated with DAPI which stains the AT regions of the DNA, washed three times for 10 min and cover slipped in Antifade Mountant with DAPI and stored at −20 °C until imaged.

3.6 Image Acquisition

Slides were imaged using the Axioimager Z1 microscope, AxiocamMR3 camera, and a 63×/1.40 oil immersion DIC M27 lens (Carl Zeiss, Canada) and the automation tool on ZEN software (Carl Zeiss, Canada). The 3D nuclear optical sectioning was done by taking 80 stacks (*Z* stacks) along the *X*, *Y*, and *Z* axis of the cell nucleus with a distance of 200 nM between each *Z* stack and then rendering the sections using a deconvolution algorithm in Axiovision 4.8 software (Carl Zeiss, Canada). The exposure time was set at 1000 ms for the Alexa Flour 647 as this time did not show any background from IgG control.

3.7 Targeted Knock Down of ATG7 Gene Modulates TGFβ_1-Induced EMT in A549 and H1975 Cells

To examine the effect of autophagy inhibition on EMT, two approaches must be examined:

First, to use pharmacological inhibitor of autophagy, such as 3-Methyladenine and Bafilomycin A1, and second, genetic manipulation of autophagy related proteins (ATGs). For chemical inhibition of autophagy, the readers are asked to refer to our previous work [65, 66]. Here we examined the efficacy of lentiviral mediated short hairpin RNA delivery to target ATGs. We therefore selected *ATG7* gene as the target gene for this goal. *ATG7* encodes the E1-like activating enzyme which is essential in the formation of autophagosome. More importantly, it activates *ATG12* for its conjugation with *ATG5* and also activation of the *ATG8* (LC3B) family proteins for their conjugation with phosphatidylethanolamine (PE) [67].

1. By using lentiviral particles carrying the shRNA against *ATG7* gene, we generated stable *ATG7* knock down clones in A549 and H1975 cells. A549 and H1975 cells were transfected with a lentiviral construct containing either a silencing shRNA against ATG7 or a non-silencing shRNA (scramble) as control at different MOIs (3 and 6).
2. Cells expressing ATG7 shRNA or scrambled control were stably selected using Puromycin selection (4 μg/ml).
3. The efficacy of our approach for downregulation of ATG7 was evaluated with Western blot using an anti-*ATG7* antibody. The Western blot analysis showed an 80–90% decrease in ATG7 protein levels, whereas in scramble clones *ATG7* protein levels were unchanged. The knock down was much more efficient with the MOI 6 in both A549 and H1975 cells.

   ATG7 knockdown (KD) A549 and H1975 cells were treated with TGFβ_1 (5 ng/ml) for 24 and 48 h according to the protocol described before. Western blotting was done on cell lysates and showed changes in the markers of EMT and autophagy. (O, P:), TGFβ_1 induces autophagy in scramble cells but not in *ATG7* KD cells. This was assessed by quantification of LC3B as it was shown by the lipidation of LC3B-II and downregulation of *ATG7*. Additionally, we could observe that in scramble cells treated with TGFβ_1, the vimentin expression was upregulated in both cells while in *ATG7* KD cells it was down regulated. We also examined the expression of E-Cadherin, as a marker of epithelial cells. In these experiments, E-Cadherin did not change significantly in either scramble or *ATG7* KD A549 and H1975 cells upon TGFβ_1 treatment. These results confirm that autophagy inhibition can abrogate the induction of EMT. Also, we detected the activation of phosphorylated Smad2 and total Smad2 in parallel with our results in the *ATG7* KD A549 and H1975 cells treated with TGFβ_1.

4 Notes

The important points that should be carefully considered and have been highlighted in the text are as following:

1. It is critically important to start the ITS starvation with proper confluence of the cells to get epithelial to mesenchymal response with the highest efficiency. The confluence of cells is dependent on the cell types and should be evaluated via growth curve of the cells in ITS starvation.
2. The starting confluence for seeding the cells for MTT assay is 2000 cells per well (in 200 μl of media).
3. The concentration of polybrene is dependent on cell type and should be obtained for each cell type.
4. MOI should be determined for each cell type.
5. The concentration of puromycin should be optimized for each cell type as the control cells without any infection have 100% cell death after 72 h of treatment.
6. Effectene Transfection Reagent volumes should be adjusted for each cell type to get at least 40–50% transfection in the cells.

In the end we emphasize that autophagy is an important physiological mechanism in mammalian cells tightly regulated in different types of tissues. Based on our recent investigations [65, 68, 69] and many other relevant studies, cellular phenotype depends on several physiological and phathophysiological events including autophagy [70] and unfolded protein response [71]. Therefore, it is highly recommended that cellular phenotype affected by autophagy in different cells/conditions response should be considered carefully in autophagy studies to prevent false positive and negative results.

Acknowledgement

SG was supported by Health Science Centre Foundation General Operating Grant and University Collaborative Research Program. SS was supported by Health Science Centre Foundation General Operating Grant and University Collaborative Research Program and Mitacs Accelerate Postdoctoral Fellowship. JA was supported by Research Manitoba studentship award.

References

1. Novikoff AB, Beaufay H, De Duve C (1956) Electron microscopy of lysosomerich fractions from rat liver. J Biophys Biochem Cytol 2 (4 Suppl):179–184
2. Madeo F, Zimmermann A, Maiuri MC, Kroemer G (2015) Essential role for autophagy in life span extension. J Clin Invest 125 (1):85–93. https://doi.org/10.1172/JCI73946
3. Mizushima N, Levine B (2010) Autophagy in mammalian development and differentiation. Nat Cell Biol 12(9):823–830. https://doi.org/10.1038/ncb0910-823
4. Deretic V, Kimura T, Timmins G, Moseley P, Chauhan S, Mandell M (2015) Immunologic manifestations of autophagy. J Clin Invest 125 (1):75–84. https://doi.org/10.1172/JCI73945
5. Jiang P, Mizushima N (2014) Autophagy and human diseases. Cell Res 24(1):69–79. https://doi.org/10.1038/cr.2013.161
6. Jiang X, Overholtzer M, Thompson CB (2015) Autophagy in cellular metabolism and cancer. J Clin Invest 125(1):47–54. https://doi.org/10.1172/jci73942
7. Kuballa P, Nolte WM, Castoreno AB, Xavier RJ (2012) Autophagy and the immune system. Annu Rev Immunol 30:611–646. https://doi.org/10.1146/annurev-immunol-020711-074948
8. Nixon RA (2013) The role of autophagy in neurodegenerative disease. Nat Med 19 (8):983–997. https://doi.org/10.1038/nm.3232
9. Kaur J, Debnath J (2015) Autophagy at the crossroads of catabolism and anabolism. Nat Rev Mol Cell Biol 16(8):461–472. https://doi.org/10.1038/nrm4024
10. Mizushima N, Komatsu M (2011) Autophagy: renovation of cells and tissues. Cell 147 (4):728–741. https://doi.org/10.1016/j.cell.2011.10.026
11. Li WW, Li J, Bao JK (2012) Microautophagy: lesser-known self-eating. Cell Mol Life Sci 69 (7):1125–1136. https://doi.org/10.1007/s00018-011-0865-5
12. Cuervo AM, Wong E (2014) Chaperone-mediated autophagy: roles in disease and aging. Cell Res 24(1):92–104. https://doi.org/10.1038/cr.2013.153
13. Zsiros V, Katz S, Doczi N, Kiss AL (2017) Autophagy is the key process in the re-establishment of the epitheloid phenotype during mesenchymal-epithelial transition (MET). Exp Cell Res 352(2):382–392. https://doi.org/10.1016/j.yexcr.2017.02.031
14. Doberstein K, Harter PN, Haberkorn U, Bretz NP, Arnold B, Carretero R, Moldenhauer G, Mittelbronn M, Altevogt P (2015) Antibody therapy to human L1CAM in a transgenic mouse model blocks local tumor growth but induces EMT. Int J Cancer 136(5): E326–E339. https://doi.org/10.1002/ijc.29222
15. Harner-Foreman N, Vadakekolathu J, Laversin SA, Mathieu MG, Reeder S, Pockley AG, Rees RC, Boocock DJ (2017) A novel spontaneous model of epithelial-mesenchymal transition (EMT) using a primary prostate cancer derived cell line demonstrating distinct stem-like characteristics. Sci Rep 7:40633. https://doi.org/10.1038/srep40633
16. Heldin CH, Vanlandewijck M, Moustakas A (2012) Regulation of EMT by TGFbeta in cancer. FEBS Lett 586(14):1959–1970. https://doi.org/10.1016/j.febslet.2012.02.037
17. Massague J (2012) TGF[beta] signalling in context. Nat Rev Mol Cell Biol 13 (10):616–630
18. Thiery JP, Acloque H, Huang RY, Nieto MA (2009) Epithelial-mesenchymal transitions in development and disease. Cell 139 (5):871–890. https://doi.org/10.1016/j.cell.2009.11.007

19. Jemal A, Siegel R, Xu J, Ward E (2010) Cancer statistics, 2010. CA Cancer J Clin 60 (5):277–300. https://doi.org/10.3322/caac.20073
20. Du B, Shim JS (2016) Targeting epithelial-mesenchymal transition (EMT) to overcome drug resistance in cancer. Molecules 21(7). https://doi.org/10.3390/molecules21070965
21. Karhadkar SS, Bova GS, Abdallah N, Dhara S, Gardner D, Maitra A, Isaacs JT, Berman DM, Beachy PA (2004) Hedgehog signalling in prostate regeneration, neoplasia and metastasis. Nature 431(7009):707–712. https://doi.org/10.1038/nature02962
22. Xie L, Law BK, Chytil AM, Brown KA, Aakre ME, Moses HL (2004) Activation of the Erk pathway is required for TGF-beta1-induced EMT in vitro. Neoplasia 6(5):603–610. https://doi.org/10.1593/neo.04241
23. Timmerman LA, Grego-Bessa J, Raya A, Bertran E, Perez-Pomares JM, Diez J, Aranda S, Palomo S, McCormick F, Izpisua-Belmonte JC, de la Pompa JL (2004) Notch promotes epithelial-mesenchymal transition during cardiac development and oncogenic transformation. Genes Dev 18(1):99–115. https://doi.org/10.1101/gad.276304
24. Shin SY, Rath O, Zebisch A, Choo SM, Kolch W, Cho KH (2010) Functional roles of multiple feedback loops in extracellular signal-regulated kinase and Wnt signaling pathways that regulate epithelial-mesenchymal transition. Cancer Res 70(17):6715–6724. https://doi.org/10.1158/0008-5472.can-10-1377
25. Xiao D, He J (2010) Epithelial mesenchymal transition and lung cancer. J Thorac Dis 2 (3):154–159. https://doi.org/10.3978/j.issn.2072-1439.2010.02.03.7
26. Bae GY, Hong SK, Park JR, Kwon OS, Kim KT, Koo J, Oh E, Cha HJ (2016) Chronic TGFbeta stimulation promotes the metastatic potential of lung cancer cells by Snail protein stabilization through integrin beta3-Akt-GSK3beta signaling. Oncotarget 7 (18):25366–25376. 10.18632/oncotarget.8295
27. Eberlein C, Rooney C, Ross SJ, Farren M, Weir HM, Barry ST (2015) E-Cadherin and EpCAM expression by NSCLC tumour cells associate with normal fibroblast activation through a pathway initiated by integrin [alpha]v[beta]6 and maintained through TGF [beta] signalling. Oncogene 34(6):704–716. https://doi.org/10.1038/onc.2013.600
28. Izumchenko E, Chang X, Michailidi C, Kagohara L, Ravi R, Paz K, Brait M, Hoque MO, Ling S, Bedi A, Sidransky D (2014) The TGFβ-miR200-MIG6 pathway orchestrates the EMT-associated kinase switch that induces resistance to EGFR inhibitors. Cancer Res 74 (14):3995–4005. https://doi.org/10.1158/0008-5472.can-14-0110
29. Koeck S, Amann A, Huber JM, Gamerith G, Hilbe W, Zwierzina H (2016) The impact of metformin and salinomycin on transforming growth factor beta-induced epithelial-to-mesenchymal transition in non-small cell lung cancer cell lines. Oncol Lett 11(4):2946–2952. https://doi.org/10.3892/ol.2016.4323
30. Yang H, Wang L, Zhao J, Chen Y, Lei Z, Liu X, Xia W, Guo L, Zhang HT (2015) TGF-beta-activated SMAD3/4 complex transcriptionally upregulates N-cadherin expression in non-small cell lung cancer. Lung Cancer 87 (3):249–257. https://doi.org/10.1016/j.lungcan.2014.12.015
31. Parvani JG, Gujrati MD, Mack MA, Schiemann WP, Z-R L (2015) Silencing β3 integrin by targeted ECO/siRNA nanoparticles inhibits EMT and metastasis of triple-negative breast cancer. Cancer Res 75(11):2316–2325. https://doi.org/10.1158/0008-5472.can-14-3485
32. Salvo E, Garasa S, Dotor J, Morales X, Peláez R, Altevogt P, Rouzaut A (2014) Combined targeting of TGF-β1 and integrin β3 impairs lymph node metastasis in a mouse model of non-small-cell lung cancer. Mol Cancer 13(1):112. https://doi.org/10.1186/1476-4598-13-112
33. Rintoul RC, Sethi T (2002) Extracellular matrix regulation of drug resistance in small-cell lung cancer. Clin Sci 102(4):417
34. Mokarram P, Albokashy M, Zarghooni M, Moosavi MA, Sepehri Z, Chen QM, Hudecki A, Sargazi A, Alizadeh J, Moghadam AR, Hashemi M, Movassagh H, Klonisch T, Owji AA, Los MJ, Ghavami S (2017) New frontiers in the treatment of colorectal cancer: autophagy and the unfolded protein response as promising targets. Autophagy 13 (5):781–819. https://doi.org/10.1080/15548627.2017.1290751
35. Macintosh RL, Timpson P, Thorburn J, Anderson KI, Thorburn A, Ryan KM (2012) Inhibition of autophagy impairs tumor cell invasion in an organotypic model. Cell Cycle 11 (10):2022–2029. https://doi.org/10.4161/cc.20424
36. Sandilands E, Serrels B, McEwan DG, Morton JP, Macagno JP, McLeod K, Stevens C, Brunton VG, Langdon WY, Vidal M, Sansom OJ, Dikic I, Wilkinson S, Frame MC (2011) Autophagic targeting of Src promotes cancer cell survival following reduced FAK signalling.

Nat Cell Biol 14(1):51–60. https://doi.org/10.1038/ncb2386

37. Grassi G, Di Caprio G, Santangelo L, Fimia GM, Cozzolino AM, Komatsu M, Ippolito G, Tripodi M, Alonzi T (2015) Autophagy regulates hepatocyte identity and epithelial-to-mesenchymal and mesenchymal-to-epithelial transitions promoting Snail degradation. Cell Death Dis 6:e1880. https://doi.org/10.1038/cddis.2015.249
38. Kenific CM, Stehbens SJ, Goldsmith J, Leidal AM, Faure N, Ye J, Wittmann T, Debnath J (2016) NBR1 enables autophagy-dependent focal adhesion turnover. J Cell Biol 212(5):577–590. https://doi.org/10.1083/jcb.201503075
39. Sharifi MN, Mowers EE, Drake LE, Collier C, Chen H, Zamora M, Mui S, Macleod KF (2016) Autophagy promotes focal adhesion disassembly and cell motility of metastatic tumor cells through the direct interaction of paxillin with LC3. Cell Rep 15(8):1660–1672. https://doi.org/10.1016/j.celrep.2016.04.065
40. Cufi S, Vazquez-Martin A, Oliveras-Ferraros C, Martin-Castillo B, Vellon L, Menendez JA (2011) Autophagy positively regulates the CD44(+) CD24(−/low) breast cancer stem-like phenotype. Cell Cycle 10(22):3871–3885. https://doi.org/10.4161/cc.10.22.17976
41. Espina V, Mariani BD, Gallagher RI, Tran K, Banks S, Wiedemann J, Huryk H, Mueller C, Adamo L, Deng J, Petricoin EF, Pastore L, Zaman S, Menezes G, Mize J, Johal J, Edmiston K, Liotta LA (2010) Malignant precursor cells pre-exist in human breast DCIS and require autophagy for survival. PLoS One 5(4):e10240. https://doi.org/10.1371/journal.pone.0010240
42. Peng YF, Shi YH, Ding ZB, Ke AW, CY G, Hui B, Zhou J, Qiu SJ, Dai Z, Fan J (2013) Autophagy inhibition suppresses pulmonary metastasis of HCC in mice via impairing anoikis resistance and colonization of HCC cells. Autophagy 9(12):2056–2068. https://doi.org/10.4161/auto.26398
43. Peng YF, Shi YH, Shen YH, Ding ZB, Ke AW, Zhou J, Qiu SJ, Fan J (2013) Promoting colonization in metastatic HCC cells by modulation of autophagy. PLoS One 8(9):e74407. https://doi.org/10.1371/journal.pone.0074407
44. Wolf J, Dewi DL, Fredebohm J, Muller-Decker K, Flechtenmacher C, Hoheisel JD, Boettcher M (2013) A mammosphere formation RNAi screen reveals that ATG4A promotes a breast cancer stem-like phenotype. Breast Cancer Res 15(6):R109. https://doi.org/10.1186/bcr3576
45. Avivar-Valderas A, Salas E, Bobrovnikova-Marjon E, Diehl JA, Nagi C, Debnath J, Aguirre-Ghiso JA (2011) PERK integrates autophagy and oxidative stress responses to promote survival during extracellular matrix detachment. Mol Cell Biol 31(17):3616–3629. https://doi.org/10.1128/mcb.05164-11
46. Chaffer CL, Weinberg RA (2011) A perspective on cancer cell metastasis. Science 331(6024):1559–1564. https://doi.org/10.1126/science.1203543
47. Fung C, Lock R, Gao S, Salas E, Debnath J (2008) Induction of autophagy during extracellular matrix detachment promotes cell survival. Mol Biol Cell 19(3):797–806. https://doi.org/10.1091/mbc.E07-10-1092
48. Kiyono K, Suzuki HI, Matsuyama H, Morishita Y, Komuro A, Kano MR, Sugimoto K, Miyazono K (2009) Autophagy is activated by TGF-beta and potentiates TGF-beta-mediated growth inhibition in human hepatocellular carcinoma cells. Cancer Res 69(23):8844–8852. https://doi.org/10.1158/0008-5472.can-08-4401
49. Kroemer G, Marino G, Levine B (2010) Autophagy and the integrated stress response. Mol Cell 40(2):280–293. https://doi.org/10.1016/j.molcel.2010.09.023
50. Li J, Yang B, Zhou Q, Wu Y, Shang D, Guo Y, Song Z, Zheng Q, Xiong J (2013) Autophagy promotes hepatocellular carcinoma cell invasion through activation of epithelial-mesenchymal transition. Carcinogenesis 34(6):1343–1351. https://doi.org/10.1093/carcin/bgt063
51. Valastyan S, Weinberg RA (2011) Tumor metastasis: molecular insights and evolving paradigms. Cell 147(2):275–292. https://doi.org/10.1016/j.cell.2011.09.024
52. Lazova R, Camp RL, Klump V, Siddiqui SF, Amaravadi RK, Pawelek JM (2012) Punctate LC3B expression is a common feature of solid tumors and associated with proliferation, metastasis, and poor outcome. Clin Cancer Res 18(2):370–379. https://doi.org/10.1158/1078-0432.ccr-11-1282
53. Zhao H, Yang M, Zhao J, Wang J, Zhang Y, Zhang Q (2013) High expression of LC3B is associated with progression and poor outcome in triple-negative breast cancer. Med Oncol 30(1):475. https://doi.org/10.1007/s12032-013-0475-1
54. Kim YH, Baek SH, Kim EK, Ha JM, Jin SY, Lee HS, Ha HK, Song SH, Kim SJ, Shin HK,

Yong J, Kim DH, Kim CD, Bae SS (2016) Uncoordinated 51-like kinase 2 signaling pathway regulates epithelial-mesenchymal transition in A549 lung cancer cells. FEBS Lett 590 (9):1365–1374. https://doi.org/10.1002/1873-3468.12172

55. Galavotti S, Bartesaghi S, Faccenda D, Shaked-Rabi M, Sanzone S, McEvoy A, Dinsdale D, Condorelli F, Brandner S, Campanella M, Grose R, Jones C, Salomoni P (2013) The autophagy-associated factors DRAM1 and p62 regulate cell migration and invasion in glioblastoma stem cells. Oncogene 32 (6):699–712. https://doi.org/10.1038/onc.2012.111

56. Nieto MA, Huang RY, Jackson RA, Thiery JP (2016) EMT: 2016. Cell 166(1):21–45. https://doi.org/10.1016/j.cell.2016.06.028

57. Ghavami S, Sharma P, Yeganeh B, Ojo OO, Jha A, Mutawe MM, Kashani HH, Los MJ, Klonisch T, Unruh H, Halayko AJ (2014) Airway mesenchymal cell death by mevalonate cascade inhibition: integration of autophagy, unfolded protein response and apoptosis focusing on Bcl2 family proteins. Biochim Biophys Acta 1843(7):1259–1271. https://doi.org/10.1016/j.bbamcr.2014.03.006

58. Ghavami S, Mutawe MM, Sharma P, Yeganeh B, McNeill KD, Klonisch T, Unruh H, Kashani HH, Schaafsma D, Los M, Halayko AJ (2011) Mevalonate cascade regulation of airway mesenchymal cell autophagy and apoptosis: a dual role for p53. PLoS One 6(1): e16523. https://doi.org/10.1371/journal.pone.0016523

59. Glogowska A, Stetefeld J, Weber E, Ghavami S, Hoang-Vu C, Klonisch T (2012) Epidermal growth factor cytoplasmic domain affects ErbB protein degradation by the lysosomal and ubiquitin-proteasome pathway in human cancer cells. Neoplasia 14(5):396–409

60. Schaafsma D, McNeill KD, Mutawe MM, Ghavami S, Unruh H, Jacques E, Laviolette M, Chakir J, Halayko AJ (2011) Simvastatin inhibits TGFbeta1-induced fibronectin in human airway fibroblasts. Respir Res 12:113. https://doi.org/10.1186/1465-9921-12-113

61. Ghavami S, Mutawe MM, Hauff K, Stelmack GL, Schaafsma D, Sharma P, McNeill KD, Hynes TS, Kung SK, Unruh H, Klonisch T, Hatch GM, Los M, Halayko AJ (2010) Statin-triggered cell death in primary human lung mesenchymal cells involves p53-PUMA and release of Smac and Omi but not cytochrome c. Biochim Biophys Acta 1803 (4):452–467. https://doi.org/10.1016/j.bbamcr.2009.12.005

62. Schaafsma D, Dueck G, Ghavami S, Kroeker A, Mutawe MM, Hauff K, FY X, McNeill KD, Unruh H, Hatch GM, Halayko AJ (2011) The mevalonate cascade as a target to suppress extracellular matrix synthesis by human airway smooth muscle. Am J Respir Cell Mol Biol 44 (3):394–403. https://doi.org/10.1165/rcmb.2010-0052OC

63. Long K, Mohan C, Anderl J, Huryn-Selvar K, Liu H, Su K, Santos M, Hsu M, Armstrong L, Ma J (2015) Analysis of autophagosome formation using lentiviral biosensors for live fluorescent cellular imaging. Methods Mol Biol 1219:157–169. https://doi.org/10.1007/978-1-4939-1661-0_12

64. Yeganeh B, Rezaei Moghadam A, Alizadeh J, Wiechec E, Alavian SM, Hashemi M, Geramizadeh B, Samali A, Bagheri Lankarani K, Post M, Peymani P, Coombs KM, Ghavami S (2015) Hepatitis B and C virus-induced hepatitis: apoptosis, autophagy, and unfolded protein response. World J Gastroenterol 21(47):13225–13239. https://doi.org/10.3748/wjg.v21.i47.13225

65. Ghavami S, Cunnington RH, Gupta S, Yeganeh B, Filomeno KL, Freed DH, Chen S, Klonisch T, Halayko AJ, Ambrose E, Singal R, Dixon IM (2015) Autophagy is a regulator of TGF-beta1-induced fibrogenesis in primary human atrial myofibroblasts. Cell Death Dis 6:e1696. https://doi.org/10.1038/cddis.2015.36

66. Yeganeh B, Ghavami S, Kroeker AL, Mahood TH, Stelmack GL, Klonisch T, Coombs KM, Halayko AJ (2015) Suppression of influenza A virus replication in human lung epithelial cells by noncytotoxic concentrations bafilomycin A1. Am J Physiol Lung Cell Mol Physiol 308 (3):L270–L286. https://doi.org/10.1152/ajplung.00011.2014

67. Gao W, Chen Z, Wang W, Stang MT (2013) E1-Like activating enzyme Atg7 is preferentially sequestered into p62 aggregates via its interaction with LC3-I. PLoS One 8(9): e73229. https://doi.org/10.1371/journal.pone.0073229

68. Gupta SS, Zeglinski MR, Rattan SG, Landry NM, Ghavami S, Wigle JT, Klonisch T, Halayko AJ, Dixon IM (2016) Inhibition of autophagy inhibits the conversion of cardiac fibroblasts to cardiac myofibroblasts. Oncotarget 7(48):78516–78531. 10.18632/oncotarget.12392

69. Ghavami S, Knight DA, Halayko AJ (2013) Autophagy is necessary for TGF-β1-induced extracellular matrix expression in idiopathic pulmonary fibrosis lung fibroblasts. Am J Respir Crit Care Med 187:A5165

70. Klionsky DJ, Abdelmohsen K, Abe A, al e (2016) Guidelines for the use and interpretation of assays for monitoring autophagy (3rd edition). Autophagy 12(1):1–222. https://doi.org/10.1080/15548627.2015.1100356

71. Shin HS, Ryu ES, ES O, Kang DH (2015) Endoplasmic reticulum stress as a novel target to ameliorate epithelial-to-mesenchymal transition and apoptosis of human peritoneal mesothelial cells. Lab Invest 95(10):1157–1173. https://doi.org/10.1038/labinvest.2015.91

Methods in Molecular Biology (2018) 1854: 105–117
DOI 10.1007/7651_2018_160

Published online: 24 May 2018

Autophagy in Zebrafish Extraocular Muscle Regeneration

Alfonso Saera-Vila, Phillip E. Kish, and Alon Kahana

Abstract

Zebrafish extraocular muscles regenerate after severe injury. Injured myocytes dedifferentiate to a mesenchymal progenitor state and reenter the cell cycle to proliferate, migrate, and redifferentiate into functional muscles. A dedifferentiation process that begins with a multinucleated syncytial myofiber filled with sarcomeres and ends with proliferating mononucleated myoblasts must include significant remodeling of the protein machinery and organelle content of the cell. It turns out that autophagy plays a key role early in this process, to degrade the sarcomeres as well as the excess nuclei of the syncytial multinucleated myofibers. Because of the robustness of the zebrafish reprogramming process, and its relative synchrony, it can serve as a useful in vivo model for studying the biology of autophagy. In this chapter, we describe the surgical muscle injury model as well as the experimental protocols for assessing and manipulating autophagy activation.

Keywords Autolysosome, Autophagy, Cell reprogramming, Dedifferentiation, Electron microscopy, EOM, Extraocular muscle, MMT, Muscle-to-mesenchymal transition, Myectomy, Regeneration, Stem cell, Zebrafish

1 Introduction

The zebrafish is a powerful model for regenerative medicine because it poses a well-known regenerative capacity. Thus, adult zebrafish can regenerate a wide range of tissues including the spinal cord [1], heart [2, 3], retina [4, 5], fins [6, 7], liver [8], and muscles [9]. Interestingly, zebrafish extraocular muscles (EOMs) can regenerate following a large myectomy [10]. To regenerate, injured muscle fibers undergo a muscle-to-mesenchymal transition which involves a dedifferentiation process and cell cycle reentry. The early steps of muscle reprogramming include both epigenetic alterations and FGF signaling to regulating proliferation of reprogrammed myoblasts [11, 12].

Macroautophagy (hereafter autophagy) is a normal cellular process that degrades unnecessary or dysfunctional components. These intracellular components, such as long-lived proteins and damaged or superfluous organelles, are engulfed within autophagosomes (double-membrane compartments) which fuse with lysosomes, generating autolysosomes, degrading their cargo [13]. Autophagy plays important roles in embryogenesis [14], degenerative conditions

[15], and cancer [16]. It has also been involved in stem cell biology [17] and cell reprogramming [18].

Autophagy is activated early on after muscle injury to remove the sarcomeric protein machinery and degrade the excess nuclei of the syncytial multinucleated myocyte, as part of the cellular reprogramming process [19]. The differential action of well-known autophagy regulators, such FGF [12] and IGF [20], in autophagy activation reveals that zebrafish EOM regeneration is an excellent model for in vivo investigation of autophagy in tissue regeneration and muscle homeostasis.

This chapter details the materials and methods required to perform EOM surgery as an injury model, detect autophagy activation (via LysoTracker staining, Western blot, transgenic lines, and electron microscopy), and utilize both pharmacologic (chloroquine treatment) and genetic (morpholino microinjection and electroporation) approaches to block autophagy.

2 Materials

Prepare all solutions using ultrapure water (prepared by purifying deionized water to attain a sensitivity of 18 MΩ cm at 25 °C) and analytical grade reagents. Prepare and store all reagents at room temperature (unless indicated otherwise). Treatment of fish is done by diluting the required solution into fish water (RO water with sodium chloride (700 μS) and sodium bicarbonate to a final pH of 7.2–7.4).

2.1 Extraocular Muscle Surgery

1. A routine stereomicroscope (e.g., Leica M50).
2. 0.05% tricaine methanesulfate (Western Chemical, Tricaine-S) prepared in fish water.
3. Thin forceps (Super Fine #5 Dumont Forceps) and microscissors (3 mm, straight blades Fine Surgical Tools FST 15000-03).

2.2 LysoTracker Staining

1. Dilute LysoTracker Red DND-99 (Thermo Fisher Scientific, L-7528) into fish water to a final concentration of 250 μM for cryosectioning or 50 nM for craniectomy (*see* **Note 1**). Prepare fresh before each use.
2. 4% paraformaldehyde (Sigma-Aldrich, 441244) in PBS pH 7.2.
3. Morse's solution. Prepare 45% formic acid (ACROS, 42375-0025) and 20% sodium citrate (R&D Systems 3161500G); mix 1:1 right before each use.

2.3 Western Blotting

Solutions for Western blots are prepared following the recommendations of *Short Protocols in Molecular Biology* 5th edition [21] with minor modifications.

1. Acrylamide/bisacrylamide, 30%/0.8%. Filter, sterilize, and store at 4 °C.
2. 4× Tris–Cl: Tris base, 1.5 M. Adjust pH to 8.8. Do not add SDS. Filter, sterilize, and store at 4 °C.
3. 4× Tris–Cl: Tris base, 0.5 M. Adjust pH to 6.8. Do not add SDS. Filter, sterilize, and store at 4 °C.
4. Ammonium persulfate (APS, Fisher, BP179): 10% solution in water.
5. *N,N,N,N′* -Tetramethyl-ethylenediamine (TEMED, Sigma, T8133). Store at 4 °C.
6. 2× loading buffer (Bio-Rad, #161-0737): 65.8 mM Tris–HCl, pH 6.8, 2.1% SDS, 26.3% (w/v) glycerol, and 0.01% bromophenol blue.
7. 10× Tris/glycine/SDS (Bio-Rad, #161-0732): 10× premixed electrophoresis buffer, contains 25 mM Tris, 192 mM glycine, 0.1% SDS, and pH 8.3 following dilution to 1× with water.
8. 10× Tris/glycine (Bio-Rad, #161-0734): Transfer buffer. Before each use prepare 1× dilution containing 20% methanol.
9. 10× Tris-buffered saline (TBS, Bio-Rad, #170-6435).
10. TBST: 1× TBS containing 0.1% of Tween 20. Store at 4 °C.
11. Primary antibodies:
    (a) Anti-γ-tubulin antibody at 1:10,000 dilution (Sigma-Aldrich, T5326).
    (b) Anti-Map1lc3a/b at 1:3,000 dilution (Novus Biologicals, NB100-2331).
    (c) Anti-Sqstm1/p62 (sequestosome 1) antibody at 1:1000 dilution (Cell Signaling Technology, 5114).
12. IgG-horseradish peroxidase conjugate secondary antibody (1:10,000 dilution):
    (a) Anti-mouse IgG, HRP-linked Antibody (Cell Signaling Technology, 7076).
    (b) Anti-rabbit IgG, HRP-linked Antibody (Cell Signaling Technology, 7074).

2.4 Transmission Electron Microscopy (TEM)

1. 4% paraformaldehyde + 1.5% glutaraldehyde (Sigma-Aldrich, 49626) in 0.1 M cacodylate buffer (pH 7.4; Hampton Research, HR2575) (*see* **Note 2**).
2. 2.5% glutaraldehyde in 0.1 M cacodylate buffer.
3. 2.5% glutaraldehyde in 7.5% disodium EDTA.
4. 1% OsO4 (ACROS Organics, ref. 191180010) in 0.1 M sodium cacodylate buffer.

5. EMbed 812 embedding resin for electron microscopy (Electron Microscopy Sciences, Hatfield, PA, USA) and copper slot grids coated with formvar.

2.5 Morpholino Microinjection and Electroporation

1. Microinjection needles: Pull microinjection capillaries (World Precision Instruments 100-F4, 1.0 mm) on a needle puller (P-97 micropipette puller; Sutter Instrument) to generate needles as you would use for injecting zebrafish embryos. Break the end of the tip with fine forceps to generate an opening (*see* **Note 3**). There is no need of polishing the needles before using them.
2. A routine stereomicroscope (e.g., Leica M50).
3. Micromanipulators (e.g., Narishige).
4. Microelectrode holder (World Precision Instruments, MPH410) attached to a pneumatic microinjector (Narishige, IM-300) with Teflon tubing to place and manipulate the microinjector needle (*see* **Note 4**).
5. Electroporator (Harvard Apparatus, Holliston, MA: BTX ECM830 electroporator) equipped with tungsten mini-electrodes.
6. Lissamine-tagged morpholinos (MOs) from Gene Tools LLC (http://www.gene-tools.com) diluted in water (*see* **Note 5**). We have successfully inhibited autophagy using the MOs listed in Table 1 (*see* **Note 6**).

2.6 Craniectomy

1. 3% agarose prepared in water (*see* **Note 7**).
2. Silicone embedding mold (PELCO® Flat Embedding Mold, 105).
3. Fluorescence stereomicroscope (Leica M205FA).

2.7 Chloroquine Treatment

1. Chloroquine (CQ, Sigma-Aldrich, C6628) was dissolved directly in fish water (*see* **Note 8**).
2. 1 L heavy wall beakers.

2.8 Transgenic Fish Lines

1. α-Actin:GFP fish: transgenic zebrafish in which the green fluorescent protein (GFP) driven by zebrafish muscle-specific actin

Table 1
Morpholino sequences used to knockdown autophagy genes

Gene	Sequence	Reference
atg5	CATCCTTGTCATCTGCCATTATCAT	[22]
becn1	CTAGAAAACCTCAAAGTCTCCATGC	[19]
becn1	TCATCCTGCAAAACACAAATGGCTT	[19]

(alpha-actin) promoter [23]. As a result, all muscles are labeled with GFP allowing their direct observation with a fluorescent microscope.

2. GFP-Lc3 fish: Transgenic fish which express GFP-tagged Lc3 under the control of the CMV (human cytomegalovirus) promoter [24]. As a result, GFP-Lc3 is expressed in most tissues allowing direct autophagy assessment using a fluorescent microscope.

3 Methods

3.1 Extraocular Muscle Surgery

1. Best results are obtained using transgenic fish with fluorescent skeletal muscles (e.g., actin::GFP). Adult zebrafish are anesthetized with Tricaine-S and placed on a paper toweling resting on the left side to perform the surgery of the right lateral rectus muscle (*see* **Note 9**) under a stereomicroscope (Fig. 1a).
2. The anterior side of the right eye is pressed with a thin probe (Fig. 1b) to rotate the eye and expose the lateral rectus muscle (Fig. 1c).
3. Fine forceps are used to separate, lift, and hold the EOM from the globe of the eye, and approximately 50% of the muscle is excised with microscissors (Fig. 1d) leaving a small remnant of the muscle attached to the ocular globe (Fig. 1e).
4. Muscle regeneration can be quantified if desired (*see* **Note 10**).

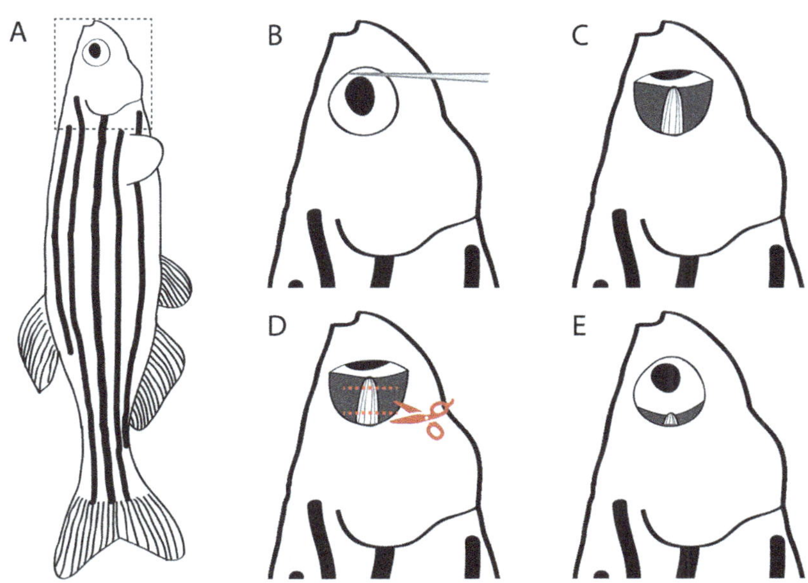

Fig. 1 Extraocular muscle surgery. Anesthetized adult zebrafish under the stereomicroscope (**a**). The anterior side of the right eye is pressed with a thin probe (**b**). Lateral rectus muscle is exposed (**c**). Approximately 50% of the muscle is excised (**d**). Remaining muscle in the insertion of the ocular globe after surgery (**e**)

3.2 LysoTracker Staining

1. Incubate adult zebrafish in LysoTracker Red DND-99 in fish water for 1 h (*see* Subheading 2.2 for appropriate LysoTracker Red DND-99 concentration).
2. Wash in fresh fish water three times for 20 min each.
3. Sacrifice treated fish by anesthesia overdose and decapitation. Zebrafish heads can be alternatively processed using the following methods (*see* **Note 11**):
   (a) Perform a craniectomy to expose the injured muscle and image using a fluorescence stereomicroscope.
   (b) Fix in 4% paraformaldehyde, decalcify in Morse's solution, and evaluate using coronal frozen sections (12 μm) with a fluorescence microscope.

3.3 Western Blotting

Western blotting is performed following the recommendations of *Short Protocols in Molecular Biology* 5th edition [21] with minor modifications as follows:

1. After a craniectomy (*see* Subheading 3.6) and surgical excision of the EOM, pool 10–15 muscles and homogenize in RIPA buffer (*see* **Note 12**). Use transgenic *α-actin:GFP* fish to visualize the muscles.
2. Sonicate (*see* **Note 13**) and centrifuge at 10,000 × *g* for 10 min at 4 °C.
3. Collect supernatant in a new tube and add loading buffer.
4. Load equal amounts of protein (20–40 μg) on SDS separating polyacrylamide gels (*see* **Note 14**) covered with a 3.9% stacking polyacrylamide gel. Separate at 130 V for 1 h.
5. Electroblot proteins into PVDF membranes (Bio-Rad, 162-0218) by wet transfer (Bio-Rad, Mini Trans-Blot® Cell) at 100 W for 1 h.
6. Block for 1 h at room temperature with 5% BSA in TBST (*see* **Note 15**).
7. Incubate overnight at 4 °C with primary antibody diluted in blocking solution.
8. Wash the membrane in TBST and incubate with IgG-horseradish peroxidase conjugate secondary antibody at room temperature for 1 h.
9. Wash the membrane in TBST.
10. Image using chemiluminescent detection.

3.4 Transmission Electron Microscopy (TEM)

1. We recommend that fish are anesthetized and then transcardially perfusion fixed in 4% PFA plus 1.5% glutaraldehyde in 0.1 M cacodylate buffer (pH 7.4) followed by decapitation (*see* **Note 16**).

2. Fix zebrafish heads overnight in 2.5% glutaraldehyde in 0.1 M cacodylate buffer.
3. Decalcify in 2.5% glutaraldehyde in 7.5% disodium EDTA for 2 days.
4. At this point, zebrafish muscles can be surgically extracted and washed in sodium cacodylate buffer.
5. Postfix for 2 h in 1% osmium tetroxide in 0.1 M sodium cacodylate buffer.
6. Wash again in sodium cacodylate buffer and dehydrate in graded ethanol followed by propylene oxide.
7. Embed with EMbed 812 orienting the EOM as desired in the embedding mold.
8. Cut thin sections (70 nm) and mount on copper slot grids coated with formvar (*see* **Note 17**).

3.5 Morpholino Microinjection and Electroporation

1. Adult zebrafish are anesthetized with Tricaine-S and placed on a paper towel under a stereomicroscope as described (Fig. 1a).
2. Press the anterior side of the right eye with a thin probe (Fig. 1b) to expose the lateral rectus muscle (Fig. 1c).
3. When the muscle is exposed, use a thin forceps or blunt probe to elevate the muscle and facilitate the microinjection (Fig. 2a). Inject the MO solution using a glass needle and a micromanipulator (*see* **Note 18**).
4. Once the MO has been microinjected into the muscle, remove the injection needle and position the positive electrode using a micromanipulator as shown in Fig. 2b.

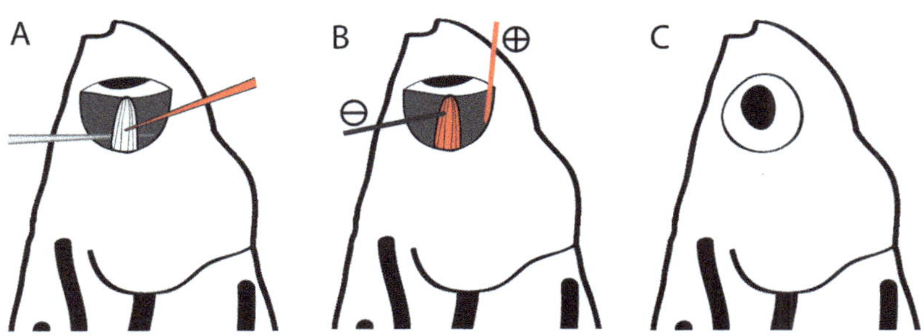

Fig. 2 Morpholino microinjection and electroporation. Lateral rectus muscle is exposed as shown in Fig. 1a–c. The muscle is elevated with the help of thin forceps or probe (gray), and the lissamine-tagged MO (red) is injected into the muscle using a glass needle (**a**). The positive electrode (red) is located inside the ocular orbit in contact with the ocular globe with the assistance of a micromanipulator. The negative electrode is manually positioned in the exposed muscle to perform the electroporating pulses (**b**). After electroporation is done, it returns to its original position (**c**)

5. Manually position the negative electrode at various locations along the muscle (Fig. 2b) performing a pulse each time (6–10 pulses at 48 V cm).
6. Remove the electrodes and allow the eye to return to its original position (Fig. 2c).
7. We routinely wait 2 h between electroporation and muscle myectomy to ensure complete diffusion of the MO inside each muscle fiber.

3.6 Craniectomy

1. Sacrifice treated fish by anesthesia overdose followed by decapitation.
2. Mount zebrafish heads in 3% agarose using the silicone embedding mold as support (Fig. 3A–C). Use a fluorescence stereomicroscope to perform the following steps and image the experimental muscle.
3. Remove the top of the skull to expose the brain (Fig. 3D), and then remove it to visualize the skull base (Fig. 3E).
4. Remove the lateral bones of the skull to allow complete visualization of the experimental muscles (Fig. 3F).
5. Image the experimental muscles using a fluorescence stereomicroscope (Fig. 3G, H).

3.7 Chloroquine Treatment

1. Perform the treatments immediately after surgery.
2. Treat up to eight fish in 1 L of CQ solution using heavy wall beakers.
3. Maintain fish at 28.5 °C under normal photoperiod.
4. Replace CQ solutions every 24 h.
5. Sacrifice fish at the required experimental time points and process as desired to assess the phenotype of autophagy inhibition.

4 Notes

1. LysoTracker Green DND-26 can also be used if the red channel is occupied. However, in our hands, the concentration of the incubation step must be increased to 1 μm for the craniectomy assay (we have not tested it in sections of fixed tissue).
2. Cacodylate buffer is used to reduce fixation artifacts causing expansion of cisternal spaces of the ER and separation around the nuclei.
3. The injection needle diameter should be big enough to allow easy penetration. Compared to embryo injecting needles, muscle injection needles require a bigger diameter. This allows

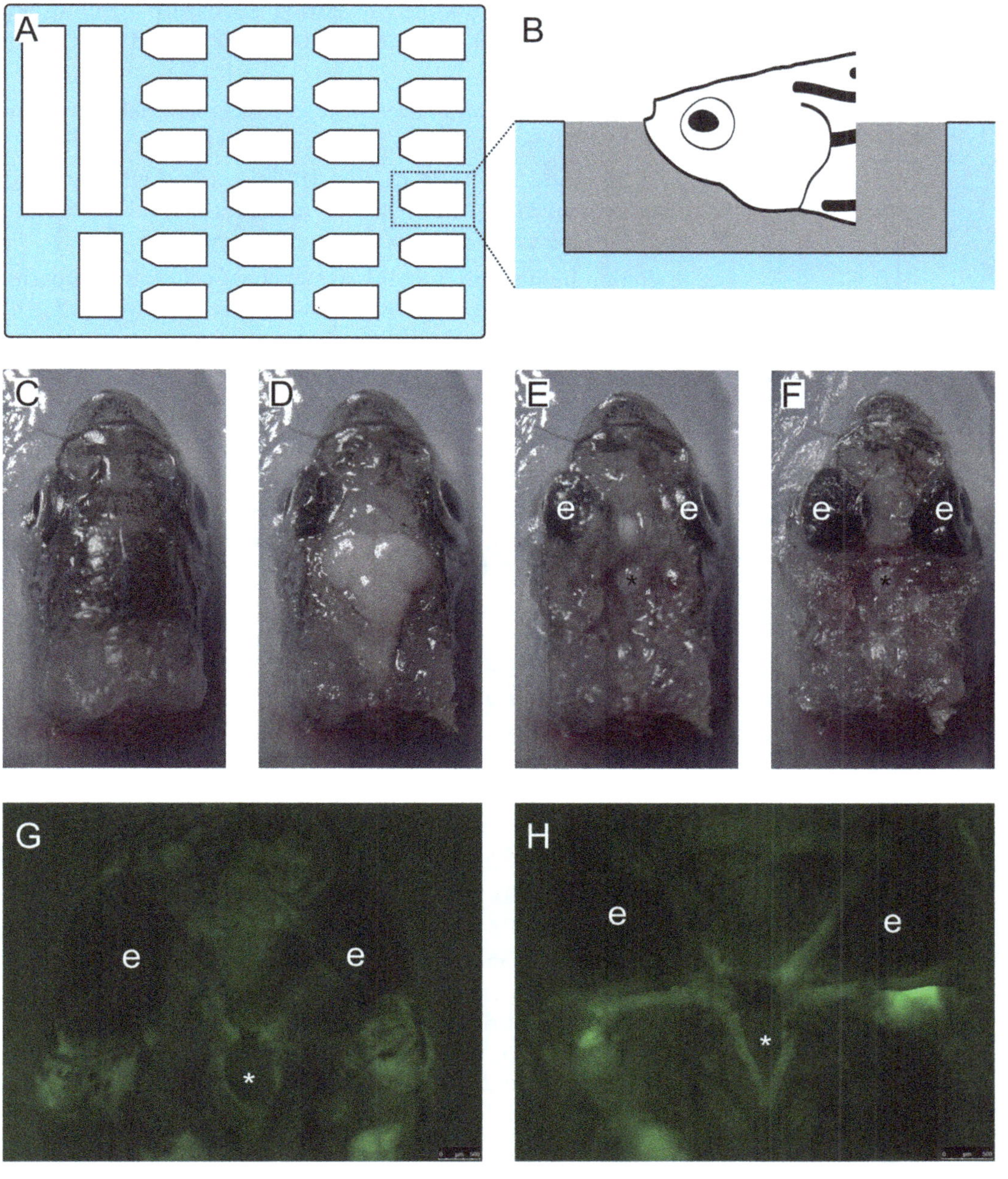

Fig. 3 Craniectomy technique. Silicone embedding mold used to mount zebrafish heads (**A**). Diagram of a zebrafish head mounted with agarose (**B**). Mounted zebrafish head (**C**). Zebrafish brain is exposed after removing the top of the skull (**D**). The bottom of the skull is exposed when the brain is removed (**E**). The lateral bones of the skull are removed to allow complete visualization of the EOMs (**F**). Fluorescent visualization of **F** showing the EOMs surrounding the pituitary at the bottom of the skull (**G**). Fluorescent visualization of **G** fully showing the EOMs (**H**). (e) Eye

muscle penetration while preventing needle tip bending and break. The optimal diameter must be determined by trial and error.

4. This setting is the same used to microinject zebrafish embryos; we refer the reader elsewhere [25, 26] for further details.
5. The lissamine tag adds to the MO the required charge for electroporation and fluorescent properties helpful to visualize the MO and assess the procedure efficiency.
6. A comprehensive list of MO sequences used against zebrafish autophagy-related genes has previously been reviewed [27, 28].
7. We recommend preparing 100 ml, store it at room temperature, and melt it before each use.
8. In our hands, a broad range of CQ concentrations (0.5–2 mM) can effectively block autophagy in zebrafish muscle. Since fish mortality is highly variable among experiments (with no clear relationship to fish or CQ batches), we routinely perform the experiments using several concentrations and select the highest one with no associated significant mortality.
9. Experimented researchers can routinely perform EOM surgeries on WT fish (or any other fish strain) using a bright light stereomicroscope. However, we recommend an initial training using α-*actin:GFP* transgenic fish, that allow fluorescent visualization of the muscles, and a fluorescent stereomicroscope. This approach will reduce the number of fish used in the training period and reduce the pain, suffering, and distress, enhancing fish welfare for the myectomyzed fish.
10. Using α-*actin:GFP* transgenic fish, regeneration can be estimated calculating the injured muscle relative growth by direct observation using craniectomy as described [10].
11. In our hand, the craniectomy approach allows LysoTracker Red DND-99 detection at much lower concentration, while the fixation and sectioning approach requires higher concentrations of LysoTracker Red DND-99 (*see* Subheading 2.2). This point should be considered when designing the experiment and would dictate the appropriate processing approach of the zebrafish heads.
12. We recommend adding protease (Roche Diagnostics Corporation, cOmplete) and phosphatase (Roche Diagnostics Corporation, PhosSTOP) inhibitors to the lysis buffer.
13. Sonication cycle must be determined empirically for each sonicator.
14. We recommend using 10-well 1.5 mm SDS polyacrylamide gels. Prepare the separating gel according to the recipe in Table 2. Use 7.5% gels for γ-tubulin and SQSTM1 and 12.5%

Table 2
Separating gel recipe

	7.5%	12.5%
30% Acrylamide/0.8% bisacrylamide	4.5 ml	7.5 ml
4× Tris–Cl, pH 8.8	4.5 ml	4.5 ml
10% SDS	0.18 ml	0.18 ml
H_2O	8.76 ml	5.76 ml
10% APS	50 μl	50 μl
TEMED	10 μl	10 μl

for LC3. When the separating gel polymerizes (~30 min), prepare the stacking gel. Stacking gel (3.9% acrylamide gel): 0.65 ml 30% acrylamide/0.8 bisacrylamide, 1.25 ml 4× Tris–Cl pH 6.8, 3.05 ml H_2O, 50 μl of 10% SDS (100×), 25 μl of 10% ammonium persulfate, and 5 μl of TEMED. Mix well and then add over the separating; put comb in. Wait until it polymerizes (~15 min).

15. 3% of BSA or nonfat dry milk have also been successfully used.
16. Fixation is better when a perfusion method is used; it prevents fixation artifacts and the damage caused by surgical extraction.
17. Cut semi-thin sections (0.5 μm), mount on glass slides, and stain with toluidine blue to find the adequate location before cutting thin sections.
18. Inject the MO in several points along the muscle length to ensure complete MO diffusion and improve efficiency of gene knockdown.

Acknowledgments

This work was funded by R01 EY022633 from the National Eye Institute (A.K.), the Alfred Taubman Medical Research Institute (A.K.), the Alliance for Vision Research (A.K.), and an unrestricted departmental grant from Research to Prevent Blindness, Inc.

References

1. Becker T, Wullimann MF, Becker CG, Bernhardt RR, Schachner M (1997) Axonal regrowth after spinal cord transection in adult zebrafish. J Comp Neurol 377(4):577–595
2. Poss KD, Wilson LG, Keating MT (2002) Heart regeneration in zebrafish. Science 298 (5601):2188–2190. https://doi.org/10.1126/science.1077857
3. Raya A, Koth CM, Buscher D, Kawakami Y, Itoh T, Raya RM, Sternik G, Tsai HJ, Rodriguez-Esteban C, Izpisua-Belmonte JC (2003) Activation of Notch signaling pathway

precedes heart regeneration in zebrafish. Proc Natl Acad Sci U S A 100(Suppl 1):11889–11895. https://doi.org/10.1073/pnas.1834204100

4. Hitchcock PF, Raymond PA (2004) The teleost retina as a model for developmental and regeneration biology. Zebrafish 1 (3):257–271. https://doi.org/10.1089/zeb.2004.1.257
5. Goldman D (2014) Muller glial cell reprogramming and retina regeneration. Nat Rev Neurosci 15(7):431–442. https://doi.org/10.1038/nrn3723
6. Poss KD, Shen J, Nechiporuk A, McMahon G, Thisse B, Thisse C, Keating MT (2000) Roles for Fgf signaling during zebrafish fin regeneration. Dev Biol 222(2):347–358. https://doi.org/10.1006/dbio.2000.9722
7. Pfefferli C, Jazwinska A (2015) The art of fin regeneration in zebrafish. Regeneration (Oxf) 2(2):72–83. https://doi.org/10.1002/reg2.33
8. Kan NG, Junghans D, Izpisua Belmonte JC (2009) Compensatory growth mechanisms regulated by BMP and FGF signaling mediate liver regeneration in zebrafish after partial hepatectomy. FASEB J 23(10):3516–3525. https://doi.org/10.1096/fj.09-131730
9. Berberoglu MA, Gallagher TL, Morrow ZT, Talbot JC, Hromowyk KJ, Tenente IM, Langenau DM, Amacher SL (2017) Satellite-like cells contribute to pax7-dependent skeletal muscle repair in adult zebrafish. Dev Biol 424 (2):162–180. https://doi.org/10.1016/j.ydbio.2017.03.004
10. Saera-Vila A, Kasprick DS, Junttila TL, Grzegorski SJ, Louie KW, Chiari EF, Kish PE, Kahana A (2015) Myocyte dedifferentiation drives extraocular muscle regeneration in adult zebrafish. Invest Ophthalmol Vis Sci 56 (8):4977–4993. https://doi.org/10.1167/iovs.14-16103
11. Louie KW, Saera-Vila A, Kish PE, Colacino JA, Kahana A (2017) Temporally distinct transcriptional regulation of myocyte dedifferentiation and Myofiber growth during muscle regeneration. BMC Genomics 18(1):854. https://doi.org/10.1186/s12864-017-4236-y
12. Saera-Vila A, Kish PE, Kahana A (2016) Fgf regulates dedifferentiation during skeletal muscle regeneration in adult zebrafish. Cell Signal 28(9):1196–1204. https://doi.org/10.1016/j.cellsig.2016.06.001
13. Boya P, Reggiori F, Codogno P (2013) Emerging regulation and functions of autophagy. Nat Cell Biol 15(7):713–720. https://doi.org/10.1038/ncb2788
14. Mizushima N, Levine B (2010) Autophagy in mammalian development and differentiation. Nat Cell Biol 12(9):823–830. https://doi.org/10.1038/ncb0910-823
15. Wong E, Cuervo AM (2010) Autophagy gone awry in neurodegenerative diseases. Nat Neurosci 13(7):805–811. https://doi.org/10.1038/nn.2575
16. Eng CH, Abraham RT (2011) The autophagy conundrum in cancer: influence of tumorigenic metabolic reprogramming. Oncogene 30 (47):4687–4696. https://doi.org/10.1038/onc.2011.220
17. Pan H, Cai N, Li M, Liu GH, Izpisua Belmonte JC (2013) Autophagic control of cell 'stemness'. EMBO Mol Med 5(3):327–331. https://doi.org/10.1002/emmm.201201999
18. Wang S, Xia P, Rehm M, Fan Z (2015) Autophagy and cell reprogramming. Cell Mol Life Sci 72(9):1699–1713. https://doi.org/10.1007/s00018-014-1829-3
19. Saera-Vila A, Kish PE, Louie KW, Grzegorski SJ, Klionsky DJ, Kahana A (2016) Autophagy regulates cytoplasmic remodeling during cell reprogramming in a zebrafish model of muscle regeneration. Autophagy 12(10):1864–1875. https://doi.org/10.1080/15548627.2016.1207015
20. Saera-Vila A, Louie KW, Sha C, Kelly RM, Kish PE, Kahana A (2018) Extraocular muscle regeneration in zebrafish requires late signals from insulin-like growth factors. PLoS One 13(2):e0192214. https://doi.org/10.1371/journal.pone.0192214
21. Ausubel FM, Brent R, Kingston RE, Moore DD, Seidman JG, Smith JA, Struhl K (2002) Short protocols in molecular biology, 5th edn. Wiley, New York
22. Hu Z, Zhang J, Zhang Q (2011) Expression pattern and functions of autophagy-related gene atg5 in zebrafish organogenesis. Autophagy 7(12):1514–1527
23. Higashijima S, Okamoto H, Ueno N, Hotta Y, Eguchi G (1997) High-frequency generation of transgenic zebrafish which reliably express GFP in whole muscles or the whole body by using promoters of zebrafish origin. Dev Biol 192(2):289–299
24. He C, Bartholomew CR, Zhou W, Klionsky DJ (2009) Assaying autophagic activity in transgenic GFP-Lc3 and GFP-Gabarap zebrafish embryos. Autophagy 5(4):520–526
25. Nüsslein-Volhard C, Dahm R (2002) Zebrafish: a practical approach, vol 975. Oxford University Press, Oxford

26. Westerfield M (2007) The zebrafish book. A guide for the laboratory use of zebrafish (Danio rerio), 5th edn. University of Oregon Press, Eugene
27. Fodor E, Sigmond T, Ari E, Lengyel K, Takacs-Vellai K, Varga M, Vellai T (2017) Methods to study autophagy in zebrafish. Methods Enzymol 588:467–496. https://doi.org/10.1016/bs.mie.2016.10.028
28. Mathai BJ, Meijer AH, Simonsen A (2017) Studying autophagy in zebrafish. Cell 6(3). https://doi.org/10.3390/cells6030021

Methods in Molecular Biology (2018) 1854: 119–130
DOI 10.1007/7651_2017_67

Published online: 10 September 2017

The Detection Techniques for Autophagy-Associated Cell Death-Related Genes and Proteins: Gene Expression Assay and Immunohistochemistry

Melek Ozturk, Ozlem Dagdeviren Ozsoylemez, and Fatma Kaya Dagistanli

Abstract

Autophagy is important in cellular homeostasis for the cell survival mechanism. Deficiency or excess of autophagy is generally related to some of diseases such as cancer and neurodegeneration. Although autophagy is a cell survival mechanism, it can mediate programmed cell death in several conditions. Autophagy-related genes (*ATGs*) regulate the autophagy and also control the crosstalk with autophagy-associated cell death and apoptosis in some condition. Various methods have been used to detect the marker genes and the proteins involved in these processes. Quantitative real-time PCR (qRT-PCR) method for monitoring the expression of genes involved in autophagy or autophagic cell death is often preferred because of its sensitivity, high efficiency potential, accurate quantification, and high-grade potential automation. The detection of the markers for autophagy-related process by immunohistochemistry in paraffin sections of various patient tissues has become a reliable method for monitoring autophagy. Here, we introduce protocols for detecting autophagy and autophagy-associated cell death in HeLa cells by using gene expression assays qRT-PCR, and also in paraffin-embedded tissue section from human biopsy material by using immunohistochemistry.

Keywords Autophagy, Autophagy-associated cell death, Gene expression assay, Immunohistochemistry

1 Introduction

Autophagy is an intracellular degredation mechanism that causes the breakdown of cytoplasmic macro-molecules and organelles that are directed to lysosomes [1]. It is essential for the maintenance of cellular homeostasis during various stress conditions, whereas uncontrolled or excessive levels of autophagy are able to induce autophagy-associated cell death. Autophagy-related genes (e.g., *ATG-5, -6, -7, -9, -12,* and *-16*) are main regulators of autophagy in mammalian development or adult tissue homeostasis [2, 3]. Apoptosis and autophagy-associated cell death have alternative cellular mechanism, whereas they can activate by the common stimulus in the same cell at the same time or sequentially [4–6]. Some of the genes such as *ATG5, Beclin-1, BNIP3, Bif-1, Bcl-2, Bcl-xL,* and *p53* provide cross-talk between apoptosis and autophagy-associated cell death [7–10]. Some of exogenous stimuli (e.g. anticancer drugs) can lead to cell death by stimulating apoptosis and/or autophagy in cancer cells [5, 7]. Stimulations of apoptosis and/or autophagy-associated cell death may provide a promising therapeutic strategy in anticancer therapies. The crosstalk regulation mechanism of apoptosis and autophagy are still not clear in both normal and disease states including cancer. Detection of certain markers of autophagy-associated proteins may provide information about the disease state. Quantitative real-time PCR is a reliable method for monitoring the expression of genes related to some markers for autophagy-associated cell death and/or apoptosis. Immunohistochemistry (IHC) is a useful method for assessing the cellular biomarkers and providing the diagnostic and prognostic information of disease status. IHC demonstrates presence and location of proteins in tissue sections. Gene expression analyses require further validation. It can be validated in tissue sections by using IHC to confirm the protein expression of selective genes [11]. The gene expression levels of the autophagy associated cell death-related genes (e.g., *ATG5, ATG12, Beclin-1, DAPk*) and the regulatory genes for both autophagic and apoptotic cell death (e.g., *Bif-1* and *BNIP3*) can be investigated by using qRT-PCR [5]. The detection of autophagy markers by IHC in paraffin sections of various patient tissues has become a reliable method for monitoring autophagy.

We introduce two protocols that can be used in autophagy and autophagy-associated cell death; (1) the gene expression assay (qRT-PCR) in HeLa cells, and (2) the immunohistochemistry in paraffin embedded tissue sections from human biopsy materials.

2 Materials

2.1 Cell Culture and Gene Expression Components

1. *Growth medium for HeLa cells*: Minimum essential medium (MEM, Sigma) containing 10% Fetal bovine serum (FBS, Gibco Lab.) and 50 mg/ml streptomycin and 100 unit/ml penicillin. MEM is sterilized by using 0.22 μl sterile filter (Millipore). MEM's pH value should be set near 7.2 using sterile 4.4% $NaHCO_3$. Store at 4 °C.
2. *Isolation of RNA:* 96–100% ethanol, 70% ethanol, 2-mercaptoethanol, PBS, and buffers. Wash buffer I (which contains guanidine isothiocyanate), wash buffer II (when using wash buffer II for the first time, add 60 ml 96% ethanol), and lysis buffer (containing guanidine isothiocyanate and 1% 2-mercaptoethanol for each purification procedure). Add 10 μl 2-mercaptoethanol for each 1 ml lysis buffer. Buffers are stored at room temperature.
3. *cDNA Synthesis:* Reverse transcription (RT) reaction mix [10 μl 2× RT buffer mix (dNTPs, random octomers, and oligo dT-

16), 1 μl 20× RT enzyme mix (MuLV reverse transcriptase and RNase inhibitor protein), and 500 pg total RNA in a final volume of 20 μl)]. Store 20× RT enzyme mix at −20 °C, and 2× RT buffer mix at 4 °C for 6 months up to use.

4. *qRT-PCR:* Master mix, probes, primers, and RT reaction plate (96 well).
5. Microcentrifuge (Hettich Zentrifugen, Universal 320R), 1.5 ml RNase-free microcentrifuge tubes, pipettors (positive-displacement, air-displacement), and vortex.
6. Qubit, real-time PCR system, Primer Express software, and GraphPad software.

2.2 Immunohistochemistry Components

1. 10% Neutral buffered formalin: 100 ml formalin (37–40% stock solution) + 900 ml distilled water, add 4 g/l NaH_2PO_4 (monobasic), and 6.5 g/l Na_2HPO_4 (dibasic/anhydrous), combine all ingredients and mix well (store at room temperature). Sodium citrate buffer (10 mM sodium citrate acid, pH 6.0): Mix 2.94 g of tri-sodium citrate in 1000 ml of distilled water, adjust pH to 6.0 with 1 N HCl, and store at 4 °C. 3% hydrogen peroxide (H_2O_2): 30% H_2O_2 (stock) 1 ml + absolute methanol 9 ml.
2. Antibodies: BECN1 *(H-300)* (Beclin-1) (sc-11427, Santa Cruz) and Autophagy APG5L (Atg5) (AP1812a, Abgent Inc.) in antibody diluent (LabVision Corp.). Immunostaining Detection Kit (Histostain Plus Bulk Kit; Invitrogen). AEC (*3-amino 9-ethyl carbazol*) Substrate Kit (Invitrogen): Add 1 drop of Reagent A to 1 ml of distilled water and mix well. Add 1 drop of Reagent B and 1 drop of Reagent C and mix well again. All are stored at 4 °C.
3. Ethanol, xylene, paraffin, Mayer's hematoxylin, and aqueous mounting medium. Microtome (Leica), light microscope (Leica), incubation oven (Elektromag), microwave, plastic staining jars, humidity chamber, positive charged (adhesive) slides, coverslips, and pipettors and its tips.

3 Methods

3.1 Gene Expression Assay: Quantitative RT-PCR [qRT-PCR]

3.1.1 Cell Line

HeLa cell line is human cervical cancer cell line (CCL-2) which can obtain from ATC (USA). HeLa cells are grown in humidity atmosphere at 37 °C and 5% CO_2 [12]. $\leq 5 \times 10^6$ adherent cells are resuspended and transferred to RNase-free centrifuge tube after adding trypsin. The growth medium is removed following centrifugation from the cells. The cells are handled according to standard practices to prevent degradation of RNA.

3.1.2 Total RNA Isolation

Total RNA is isolated as described in the protocol (AmbionPure-Link RNA Mini Kit).

1. The cells are resuspended in 600 μl lysis buffer with 2-mercaptoethanol in centrifuge tube (*see* **Note 1**). Vortex until the cell pellet is dispersed and the cells appear lysed. Transfer the lysate to a new RNase-free homogenization tube. Perform manual homogenization. Spin the sample at 12,000 × *g* on a microcentrifuge for 2 min at room temperature. Transfer the supernatant to a new RNase-free tube.
2. Add 70% ethanol to equal volume of cell homogenate, and mix the sample by using the vortex machine. Add sample to spin cartridge (with the collection tube) to bind RNA membrane and spin at 12,000 × *g* for 15 s at room temperature. Discard the flow-through and reinsert the spin cartridge into the same collection tube.
3. Add 700 μl Wash Buffer I to the spin cartridge. Spin at 12,000 × *g* for 15 s. Discard the collection tube containing the flow-through solution. Insert the spin cartridge into the new collection tube.
4. Add 500 μl wash buffer 2 with ethanol (see Note 2) to the spin cartridge. Spin at 12,000 × *g* for 15 s at room temperature. Discard the collection tube containing the flow-through solution. Repeat this step once. Spin at 12,000 × *g* for 1–2 min at room temperature.
5. Discard the collection tube and transfer the spin cartridge with RNA bounded membrane, to sterile 1.5 RNase-free microcentrifuge tube (included in kit). Add 100 μl nuclease-free water (included in kit) to the center of the spin cartridge and incubate for 1 min. Centrifuge for 2 min at ≥12,000 × *g* to elute RNA into recovery tube (*see* **Note 3**).
6. Discard the spin cartridge. Total RNA in collection tube is determined via Qubit (Invitrogen, Qubit 2.0 Fluorometer) (*see* **Note 4**).

3.1.3 Two-Step RT-PCR

Gene quantitation assays using TaqMan Gene Expression Master Mix and TaqMan. Gene expression assay are performed by the following two steps: (1) in the reverse transcription (RT) step, cDNA is reverse transcribed from RNA, and (2) in the PCR step, PCR products are quantitatively synthesized from cDNA samples using the TaqMan Gene Expression Master Mix.

3.1.4 cDNA Synthesis

For each group, eight wells are used. Each well contains RT reaction mix (Applied Biosystems, High Capacity RNA-to-cDNA Kit). It is important to keep the reactions on ice (*see* **Note 5**). The mix is incubated at 37 °C for 60 min. Reaction is stopped by heating at 95 °C for 5 min and obtained cDNAs are stored at 4 °C (cDNA is determined via Qubit 2.0 Fluorometer, Invitrogen). The cDNA is used for qRT-PCR (*see* **Note 6**).

3.1.5 qRT-PCR

1. For each group, 400 μl nuclease free water, 500 μl Master Mix (e.g., Applied Biosystem, TaqMan Gene, expression Master Mix), and 100 μl cDNA are added in the well of the plate that include lyophilized primers and probes (TaqMan Array 96-Well Fast Plate, Custom Format 48, Applied Biosystems) (each mRNA is repeated four times).
2. The samples are established qRT-PCR appliance (e.g., Applied Biosystems, 7500 Fast Real-Time PCR Systems) as described in the following table.

Predenaturation	PCR amplification
1 Cycle:	*40 Cycles:*
At 50 °C for 2 min	At 95 °C for 15 s
At 95 °C for 10 min	At 60 °C for 1 min

3. The mRNA levels of the autophagic cell death-related genes (e.g., *ATG5, ATG12, Beclin-1 and DAPk*), and the common regulatory genes of autophagy and apoptotic cell death (e.g., *Bif-1 and BNIP-3*), and also the apoptotic genes (e.g., *Bcl-xL, PUMA, NOXA, Caspase-3,-8,-9, Fas, FADD, TRADD, TNFR1* and *TRAF2*) are determined by qRT-PCR. GADPH, 18S, and actin β are used for housekeeping genes [5].

Master Mix: Master mix (TaqMan Gene Expression Master Mix) includes AmpliTaq Gold® DNA polymerase, UP (Ultra Pure), uracil-DNA glycosylase (UDG), deoxyribonucleotide triphosphates (dNTPs) with deoxyuridine triphosphate (dUTP), ROX™ Passive Reference, and buffer components (*see* **Note 7**).

Probes: Probes (The TaqMan® MGB Probes) consists of a classic target-specific oligonucleotide and other three inserts: First, a 5′ end linked reporter dye (for example, 6FAM™ dye) and second, a minor Groove binder (MGB), which increases the melting temperature (Tm) [13], shorter and more accurate probes. Finally, 3′ end linked non-fluorescent quencher (NFQ) to prevent smoldering of the dye which causes high background signal and poor quantitation. Custom designing of primers and probes in real-time qPCR assays involves some details that is described in the following guides by the producer.

Custom Primer Design: Determining the target template and amplicon, designing Custom TaqMan® Probes and Sequence Detection Primers (CTMP-SDP), quantitating CTMP-SDP, determining optimal CTMP-SDP concentrations, and performing real-time quantitative PCR. Primers can be designed using Primer Express software (in our study the primers were designed by using this step (5) (ID: 08062012, AB Applied Biosystem, USA) (*see* **Note 8**).

Custom Probes Design: Probes can be designed using Primer Express software (*see* **Note 9**).

3.1.6 RT-Reaction

The reaction mixture is incubated at 50 °C for 2 min for optimal UDG enzyme activity, then 1 min at 60 °C for UP enzyme activity and template denaturation. 40 cycles of PCR are performed, each cycle consisting of 15 minutes at 95 °C and 1 minute at 60 °C. Fluorescence data is acquired after each annealing and extension step. Results are expressed using the comparative threshold (Ct) method. FC (fold change ≥2) value is calculated by using a target gene expression ratio [5]. Endogenous reference gene (housekeeping gene) expression level should not differ between samples. Comparing the CT value of a target gene with those of the reference gene allows normalization of the expression level of the target gene to the amount of input RNA or cDNA [14]. The expression level of the reference gene must not vary in different states of the same tissue or cell line. (*see* **Note 10**).

3.1.7 Data Analysis

One way ANOVA can be used to determine the statistical difference between experimental groups, depending on time and treatments. Ct values that are obtained from qRT-PCR are used for determining the relative expression levels of the target genes [5] (GraphPad Prism version 4.00, GraphPad Software, San Diego California, USA).

3.1.8 Rn and Δ Rn Values

In order to achieve normalized reporter (*Rn*) value for each tube, emission intensity of a passive reference dye called ROX should be proportioned against the emission intensity of the reporter dye. *Rn*+ is derived from the *Rn* value of a reaction containing all components including the template. *Rn*− is derived from the *Rn* value of an unreacted sample. The best to obtain this value is to form a reaction without any template. ΔRn is the magnitude of the signal generated by the given set of PCR conditions. The ΔRn is the difference between the Rn at the end point (Rn+) and at the starting point (Rn−) [15]. The equation below is symbolizing the relationship of positive and negative normalized reporters: $\Delta Rn = (Rn+)-(Rn-)$:

$$R_n^+ = \frac{\text{Emission Intensity of Reporter}}{\text{Emission Intensity of Passive Reference}} \quad \textit{PCR with template},$$

$$R_n^- = \frac{\text{Emission Intensity of Reporter}}{\text{Emission Intensity of Passive Reference}} \quad \textit{PCR without}$$

template or early cycles of a real – time reaction.

3.1.9 CT (Threshold Cycle)

The threshold cycle (CT) can be determined from *Rn* versus cycle number graphs where the first detectable ΔRn peak occurs. The intersection point of the threshold with the ΔRn which apparently is an expression of exponential amplification of PCR product defines the CT in real-time PCR assays [16]. It is suggested to set the threshold between the background signal and the exponential amplification curve, or to set the average standard deviation of *Rn* for the early cycles.

3.1.10 Quantitation of cDNA Relative to a Calibrator Sample

In order to determine the gene expression quantitatively from cDNA in a relative manner, we need a calibrator sample which represents a physiological reference. The calibrator samples should be chosen from a special or an untreated group carefully to compare gene expression levels against an inspected group or a treated group for each different experiment.

3.2 Immunohistochemistry

3.2.1 Tissue Preparation and Immunostaining

The tissue specimens are fixed in 10% neutral buffered formalin and embedded in paraffin, and then cut into 5 μm thick sections by microtome. The sections are placed onto positive charged slides, deparaffinized in xylene, and rehydrated in graded alcohol [6].

The protocol is according to immunoperoxidase technique by using *Histostain Plus Bulk Kit* (Invitrogen CA, USA). The tissue should not be dried out during all the following steps.

1. Deparaffinize tissue sections in xylene using two changes for 20 min each, at room temperature. Hydrate sections gradually through graded alcohols; 100% ethanol twice for 5 min each, 95 and 70% alcohol for 5 min each, wash slides in deionized H_2O twice for 2 min.
2. *Antigen Retrieval:* Place slides in microwave-resistant plastic staining jar containing 0.01 M sodium citrate buffer, pH 6.0 (*see* **Note 11**). Operate the microwave three times for 5 min on medium power (~600 W). Allow slides to cool in the buffer at room temperature for 20 min. Wash slides in deionized H_2O for 3 min each. Wash slides three times for 3 min in PBS.
3. *Quench Endogenous Peroxidase:* 3% hydrogen peroxidase in methanol to block endogenous enzyme activities for 10 min. Wash slides three times for 3 min in PBS.
4. *Serum Blocking:* Immediately apply 100 μl of serum blocking solution (10% goat non-immune serum; included in kit) directly on each section. Incubate for 15 min (*see* **Note 12**). Gently drain or blot off solution around the sections. *Do not rinse.*
5. Dilute primary antibody [e.g., BECN1 (*H-300*), 1:100 and APG5, 1:75 in antibody diluent]. Apply enough of the primary antibody to completely cover the tissue on each section. Incubate overnight in humidity chamber at 4 °C. Wash slides three times for 3 min in PBS (*see* **Note 13**).

6. *Secondary Antibody:* Add two drops of biotinylated secondary antibody (included in kit) to each slide and incubate for 15 min (*see* **Note 14**). Wash slides three times for 3 min in PBS.
7. *Enzyme Conjugate:* Add two drops of horse radish peroxidase (HRP)–streptavidin complex to each slide and incubate for 15 min. Wash slides three times for 3 min in PBS. Add 100 μl of AEC substrate–chromogen mixture working solution to each slide (*see* **Note 15**). Incubate sections for 5–15 min at room temperature. Color development should be monitored under a light microscope.
8. Wash slides in distilled water. Counterstain slides with Mayer's hematoxylin and incubate for 45 s. Wash slides with tap water and incubate slides in PBS (pH 7.4) for 30 s until color changes blue. Rinse well with distilled water. Then slides should be mounted with aqueous mounting solution with coverslip (*see* **Note 16**). The stained slides are ready to examine under the light microscope (Fig. 1).

Staining specificity controls: Positive and negative controls are the most valuable tool to monitor the specificity and sensitivity for IHC [17]. Breast cancer tissue sections can be used as positive control. For negative controls, distilled water is used instead of primary antibody.

3.2.2 Semi-Quantitative Analysis

The immunostaining intensity is scored as: no (−), weak (+), moderate (++), and strong (+++). A histological score (*H* score) is derived for each slide by summing the percentage of stained cells at each intensity category and multiplying that value by weighted intensity of the staining, according to the formula: *H* Score ΣPi $(i + 1)$ (Pi; corresponding percentage of cells, i; intensity of score). 5 or 10 randomly selected areas in each slide are evaluated under a

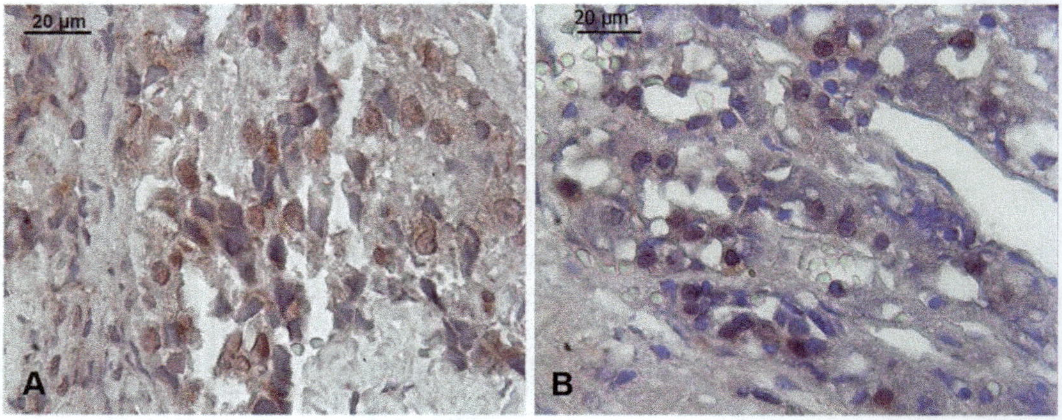

Fig. 1 Immunolocalization of Beclin-1 (**a**) and Atg5 (**b**) in the human pituitary tissue. Beclin-1 and Atg5 positive signals are seen in red color in both the cytoplasm and nucleus of the cell (immunostaining: streptavidin–biotin peroxidase; nucleus: hematoxylin)

light microscope. The percentage of cells at the different intensities is determined by two researchers blinded to the study. The intra- and inter-individual coefficients of variation are 10 and 12%, respectively, for the *H* score evaluation [18].

4 Notes

1. Lysis buffer containing 1% 2-mercaptoethanol is prepared fresh. 10 μl 2-mercaptoethanol added for every 1 ml lysis buffer.
2. When using wash buffer II for the first time, add 60 ml of 96–100% ethanol.
3. If the expected RNA yield is 100 μg, perform 3 sequential elutions of 100 μl each. Collect elutes in a single tube.
4. Use the purified RNA for downstream applications or store RNA on ice for immediate use or at −80 °C until use.
5. Centrifuge the tubes to spin down the contents and to eliminate any air bubbles.
6. Store the cDNA at 4 °C if used within 24 hours, store at −20 °C for a long time.
7. In the course of PCR amplification, the DNA polymerase, UP enzyme reduces bacterial DNA introduced from the host organism owing to the purification process of the enzyme under production. Uracil-DNA glycosylase treatment prevents the reamplification of contaminating molecules in PCR products by substituting dUPT for dTTP. During subsequent cycle only target DNA is amplified.
8. Choose the primers after selecting the probe. Design the primers as close as possible to the probe without overlapping the probe. Keep the G-C content in the 20–80% range. Avoid runs of an identical nucleotide, especially for guanine, where runs of four or more Gs should be avoided. The Tm of each primer should be 58–60 °C. The five nucleotides at the 3′ end of each primer should have no more than two G and/or C bases.
9. The probes are designed. Follow these guidelines when designing probes: keep the G-C content in the 20–80% range; avoid runs of an identical nucleotide, especially for guanine, where runs of four or more Gs should be avoided; do not put a G on the 5′ end; select the strand that gives the probe more Cs than Gs; the probe melting temperature (Tm) should be 68–70 °C when using Primer Express software; use Primer Express software 1.5a or later when designing TaqMan MGB probes. For duplex assays, both probes should be on the same strand.

10. There are several other problems that may be encountered in real-time PCR experiments. Amplification can be altered unusually due to the composition of the RNA extract what is reported to contain inhibitors. Although free from the inhibitors, amplicon length and primer quality affect the fluorescence generated amplification curves as much as they shift the background fluorescence levels. In conclusion, efficiency of the PCR oversees quantification of the genes where it mainly depends on the competence of the fluorescence history, primer and probe properties, and quality of the reaction components [19].
11. Formalin fixation forms methylene bridges between proteins that mask antigenic sites of primary antibodies in tissue; therefore formalin-fixed tissues require an antigen retrieval step before the staining. The masked epitopes can be retrieved by using heat induced (HI) or proteolytic induced (PI) epitope retrieval. HI epitope retrieval technique is often performed using microwave ovens, and the most commonly used buffer is 10 mM sodium citrate, pH 6.0 [20]. Proteinase-K, trypsin, or pepsin is used for PI epitope retrieval. The antigen retrieval technique is preferred depending on the tissue, the fixation method, and/or primary antibody, and it must be optimized.
12. Serum blocking solution is used to eliminate nonspecific background.
13. IHC staining for the detection of proteins for autophagy and/or autophagy-associated cell death proteins (e.g., Atg5, Beclin-1) is performed [6].
14. Secondary antibodies are specific for the isotype and the species of the primary antibody. For example, a rabbit primary antibody binds an anti-rabbit secondary antibody.
15. AEC solution is prepared fresh, and it is kept away from light.
16. Consider the following for inadequate results in immunostaining;

    Possible causes of no staining: Specimen was inadequately fixed and/or processed; modify the fixation procedures and use different antigen retrieval methods. The primary antibody and the secondary antibody are not compatible; use appropriate antibodies (e.g., primary is raised in rabbit, use anti-rabbit secondary). The protein is not presented in tissue of interest; use the control tissue sections containing the antigen.

    Possible causes of weak staining: Too much rinse buffer left on slides causing excessive dilution of reagents. Concentrations of antibody or linker incubation times are insufficient. Deparaffinization may be incomplete; use fresh xylene and deparaffinize longer.

Possible causes of high background: Blocking of nonspecific binding might be insufficient; increase blocking incubation period or change blocking agent. The primary antibody concentration may be too high; try to find optimal concentrations for antibody and incubate at 4 °C. Substrate for enzymatic detection is applied too much; dilute substrate and reduce substrate incubation time.

Acknowledgment

This study was supported by the Research Fund of the University of Istanbul, Turkey, Project numbers: 17492, 38781 and 39211.

References

1. Ohsumi Y (2001) Molecular dissection of autophagy: two ubiquitin-like systems. Nat Rev Mol Cell Biol 2:211–216
2. Mizushima N, Levine B, Cuervo AM, Klionsky DJ (2008) Autophagy fights disease through cellular self-digestion. Nature 451:1069–1075
3. Lin L, Baehrecke EH (2015) Autophagy, cell death and cancer. Mol Cell Oncol 2:e985913
4. Radogna F, Dicato M, Diederich M (2015) Cancer-type specific crosstalk between autophagy, necroptosis and apoptosis as a pharmacological target. Biochem Pharmacol 94:1–11
5. Ozsoylemez O, Ozturk M, Sutlupinar N, Kayacan S, Tuncdemir M, Ozcan G (2016) The effects of Colchicum baytopiorum on regulatory genes of apoptotic and autophagic cell death in HeLa cells. Curr Pharm Biotechnol 17:1369–1376
6. Dagistanli FK, Ozkaya HM, Kucukyoruk B, Biceroglu H, Metin D, Gazioglu N et al (2016) Preoperative somatostatin analogue treatment might trigger apoptosis and autophagy in tumor tissues of patients with acromegaly: a pilot study. Exp Clin Endocrinol Diabetes. [Epub ahead of print]
7. Rubinstein AD, Kimchi A (2012) Life in the balance a mechanistic view of the crosstalk between autophagy and apoptosis. J Cell Sci 125:5259–5268
8. Kang R, Zeh HJ, Lotze MT, Tang D (2011) The Beclin-1 network regulates autophagy and apoptosis. Cell Death Differ 18:571–580
9. Mathew R, Karantza-Wadsworth V, White E (2007) Role of autophagy in cancer. Nat Rev Cancer 7:961–967
10. Gordy C, He YW (2012) The crosstalk between autophagy and apoptosis: where does this lead? Protein Cell 3:17–27
11. Ip JC, Pang TC, Glover AR, Soon P, Zhao JT, Clarke S et al (2015) Immunohistochemical validation of overexpressed genes identified by global expression microarrays in adenocortical carcinoma reveals potential predictive and prognostic biomarkers. Oncologist 20:247–256
12. Freshney RI (2000) Cell lines culture of animal cells: a manual of basic technique. Wiley-Liss, New York
13. Pestana EA, Belak S, Diallo A, Crowther JR, Viljoen GJ (2010) Early, rapid and sensitive veterinary molecular diagnostics-real time PCR applications. Springer, New York
14. VanGuilder HD, Vrana KE, Freeman WM (2008) Twenty-five years of quantitative PCR for gene expression analysis. Biotechniques 44:619–626
15. Yajima T, Yagihashi A, Kameshima H, Kobayashi D, Furuya D, Hirata K et al (1998) Quantitative reverse transcription-PCR assay of the RNA component of human telomerase using the TaqMan fluorogenic detection system. Clin Chem 44:2441–2445
16. Letertre C, Perelle S, Dilasser F, Arar K, Fach P (2003) Evaluation of the performance of LNA and MGB probes in 5'-nuclease PCR assays. Mol Cell Probes 17:307–11
17. Torlakovic EE, Francis G, Garratt J, Gilks B, Hyjek E, Ibrahim M et al (2014) International ad hoc expert panel. Standardization of negative controls in diagnostic

immunohistochemistry: recommendations from the international ad hoc expert panel. Appl Immunohistochem Mol Morphol 22:241–252

18. Kaya-Dagistanli F, Tanriverdi G, Altinok A, Ozyazgan S, Ozturk M (2013) The effects of alpha lipoic acid on liver cell damages and apoptosis induced by polyunsaturated fatty acids. Food Chem Toxicol 53:84–93

19. Fraga D, Meulia T, Fenster S (2008) Real-time PCR. Current protocols essential laboratory techniques. Doi: 10.1002/9780470089941.et1003s00

20. Vinod KR, Jones D, Udupa V (2016) A simple and effective heat induced antigen retrieval method. MethodsX 3:315–319

Methods in Molecular Biology (2018) 1854: 131–146
DOI 10.1007/7651_2017_83

Published online: 12 October 2017

Cloning of Autophagy-Related MicroRNAs

Deniz Gulfem Ozturk, Muhammed Kocak, and Devrim Gozuacik

Abstract

Autophagy is a cellular survival pathway that is necessary for the degradation of cellular constituents such as long-lived proteins and damaged organelles. Conditions resulting in cellular stress such as starvation or hypoxia might activate autophagy. Being at the crossroads of various cellular response pathways, dysregulation of autophagy might result in pathological states including cancer and neurodegenerative diseases. Autophagy has also been shown to participate in stemness. MicroRNAs were introduced as novel regulators of autophagy, and accumulating results underlined the fact that they constituted an important layer of biological control mechanism on the autophagic activity.

MicroRNAs are protein noncoding small RNAs that control cellular levels of transcripts and proteins through posttrancriptional mechanisms. Novel miRNAs in human and mouse genomes are yet to be identified. Considering the emerging role of autophagy in health and disease, identification of novel autophagy-regulating miRNAs and determination of relations between miRNA expression and physiological and pathological conditions might contribute to a better understanding of mechanisms governing health and disease. High-throughput techniques were developed for miRNA profiling, yet for a thorough characterization and miRNA target determination, miRNA cloning remains as an important step. Here, we describe a modified miRNA cloning method for the characterization of novel autophagy-regulating miRNAs.

Keywords Autophagy, Cloning, Identification, miRNA, Regulation

1 Introduction

Macroautophagy (hereafter referred as autophagy) is an evolutionarily conserved catabolic and degradative provess that is active under basal conditions in alleukaryotic cell types, allowing intracellular recycling and homeostasis. Autophagy is one of the two major catabolic pathways in cells: The ubiquitin–proteasome system (UPS) degrades short-lived and mostly soluble proteins whereas basal autophagy pathways are responsible for the degradation of misfolded and aggregate-prone proteins or damaged organelles. Autophagy can be rapidly upregulated in response to a spectrum of stress conditions, including nutrient, energy or growth factor deprivation, hypoxia, endoplasmic reticulum stress, intracellular pathogens and toxins. Thus it represents an important system for quality control in cellular homeostasis and stress response [1]. Activation of autophagic degradation under stress conditions provides energy and nutrients and limits oxidative stress, helping cells to survive unfavorable periods and avoid demise. Nevertheless, uncontrolled and excessive activation of autophagy may damage cells and even lead to a caspase-independent, non-apoptotic cell death type that is called as "autophagic cell death" [2]. Therefore, autophagic activity should be under strict control in cells.

Alterations in autophagy are associated with several health problems, including neurodegenerative, cardiovascular and infectious diseases, as well as cancer. Accumulating evidence indicate that autophagy also participates in the regulation of cell proliferation, differentiation, somatic reprogramming and stemness in various cell types [3]. For example, autophagic activity is shown to be constitutively high in human mesenchymal stem cells (hMSC), hematopoietic stem cells (HSC), dermal stem cells (DSC), and epidermal stem cells (Epi-SCs), thereby promoting their maintenance [4, 5]. Once cells differentiated, autophagic activity is downregulated to a basal level. In contrast, autophagy is upregulated during differentiation of neural and cardiac stem cells (NSC and CSC) [6] and early differentiation of human embryonic stem cells (ESC) [7]. Autophagy also functions in somatic reprogramming and it was introduced as a positive regulator of induced pluripotency through inhibition of senescence and apoptosis [8].

microRNAs are endogenously expressed noncoding small RNAs that control cellular levels of transcripts and proteins through recognition of miRNA-response elements (MREs) that are usually found in the 3′ UTR of target messanger RNAs [9] (*see* Fig. 1). Being endogenous regulators of gene expression, several miRNAs were recently shown to play a role in the regulation of cellular pathways. Indeed, independent studies demonstrated that, core autophagy-related genes (*ATG* genes) and upstream mediators were targeted by microRNAs, revealing the presence of a novel and intricate miRNA network that is tightly regulating autophagy under physiological conditions. Moreover, dysregulation of miRNA expression was reported under various pathological conditions, including cancer, neurodegenerative diseases, and cardiac and metabolic disorders [10]. Most autophagy-related miRNAs were shown to be upregulated or downregulated in response to autophagy-inducing stress signals. In order to achieve a dynamic and context-dependent regulation, stress responsiveness may be an important property of autophagy modulation by miRNAs.

In the literature, a number of different approaches were used in order to profile miRNAs. In addition to Northern hybridization which requires large amounts of total RNA (not suitable for low abundance miRNAs), high-throughput profiling techniques were developed. These techniques that require very small amounts of RNA are suitable for miRNA analyses including miRNA microarrays [11] and quantitative reverse transcription-mediated polymerase chain reaction (qRT-PCR) [12]. A major limitation of these

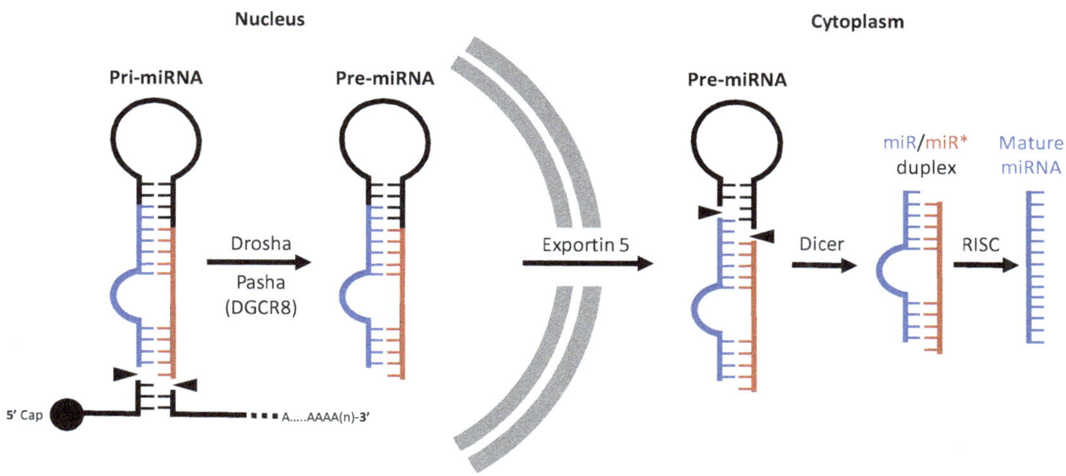

Fig. 1 The long primary miRNA transcript (pri-miRNA) are transcribed from miRNA genes in an RNA polymerase II (Pol II)-dependent manner, and form a local hairpin structure called pri-miRNA. Following transcription, evolutionarily conserved mechanisms of human miRNA biogenesis give rise to mature miRNAs following nuclear and subsequent cytoplasmic cleavage events: Following transcription, the nuclear RNase III Drosha cleaves and forms pre-miRNAs with a ~60–100 nt hairpin structures. Pre-miRNAs are then transported into the cytoplasm through exportin-5-dependent nuclear export. In the cytoplasm, pre-miRNAs are subject to a second processing event that is catalyzed by Dicer enzymes. Double-stranded ~22 ntRNAs are then produced. They consist of a mature miRNA guide strand and a miRNA* passenger strand. The mature miRNA guide strand is chosen by the RNA-induced silencing complex (RISC). MiRNAs guide the RISC to mRNA targets and eventually lead to gene silencing through their degradation or translation inhibition

techniques is that they can only detect known miRNAs. But additional new miRNAs are present in the human and mouse genomes, and they are yet to be identified.

Small RNA cloning is a powerful and sensitive approach that identifies novel miRNAs. Moreover, miRNA cloning might allow determination of the relative abundance of cloned miRNAs through establishment of the cloning frequency of an individual miRNA in a library. Original cloning protocols were developed by Ambros, Bartel, and Tuschl groups [13–15]. But several additional protocols and their modifications were adapted independently as alternative miRNA cloning methods. Most cloning techniques use small RNAs that are 200 bases or smaller as a starting material. Although by this way a significant miRNA enrichment can be achieved, possibility of a contamination of purified RNAs with piRNAs should be considered. Moreover, according to published results, nearly 50% of identified clones might correspond to degradation products of rRNAs, tRNAs, or snRNAs [16].

The miRNA cloning protocol in this chapter was based on methods from both Mello Lab [17] and Fu et al. [18]. Basically, small RNAs isolated from total RNA of cells are first fractionated in order to obtain 18–24 nt miRNAs. MiRNAs are then polyadenylated using RNA polymerase. An RNA oligo adapter is added to 5′

ends of the polyadenylated small RNAs by using an RNA ligase. Following reverse transcription with an RT primer, the resultant small RNA-derived cDNAs are then amplified usinga pair of 5′ and the 3′ linker primers. The amplified cDNAs are cloned in TA cloning vectors for the ease of sequencing and identification of autophagy-related miRNAs. (*See* Fig. 2).

2 Materials

2.1 Autophagy Inducers

1. Torin-1: 1 mM Torin-1 (Tocris) dissolved in DMSO (Sigma). Stable at −20 °C for at least 3 months.
2. Starvation media: Earle's Balanced Salt solution (EBSS; Biological Industries).

2.2 Solutions and Reagents

1. RNA Gel loading buffer II (Ambion) (*see* Note 1).
2. 10 bp DNA ladder (Thermo).
3. M15 100 bp DNA ladder (SibEnzyme).
4. Glycogen (Santa Cruz) (20 mg/ml in nuclease-free water).
5. Ethidium Bromide (10 mg/ml).
6. Gel Elution Buffer.

   0.5 M Ammonium Acetate.
   10 mM Magnesium Acetate.
   1 mM EDTA (pH 8.0).
   0.2% (w/v) SDS.
7. Tris–EDTA Buffer.

   10 mM Tris–HCl.
   1 mM EDTA.
8. Phenol–Chloroform–Isoamyl Alcohol (25:24:1) (Thermo).
9. IPTG (100 mM, soluble in water).
10. X-Gal (20 mg/ml, soluble in DMSO).
11. Diethyl pyrocarbonate treated H_2O: Add 1 ml of fresh DEPC to 1 l of H_2O. Stir DEPC-H_2O overnight, then autoclave for 20 min to inactivate the remaining DEPC.
12. Isopropanol (Sigma).
13. Ethanol (Sigma).
14. Sodium Acetate (3 M, pH 5.2).

2.3 Small RNA Isolation

miRVana miRNA Isolation kit (Thermo).

2.4 Denaturing UREA PAGE Gel

1. Urea ultra pure (Thermo).
2. 20× Tris–borate–EDTA (TBE).

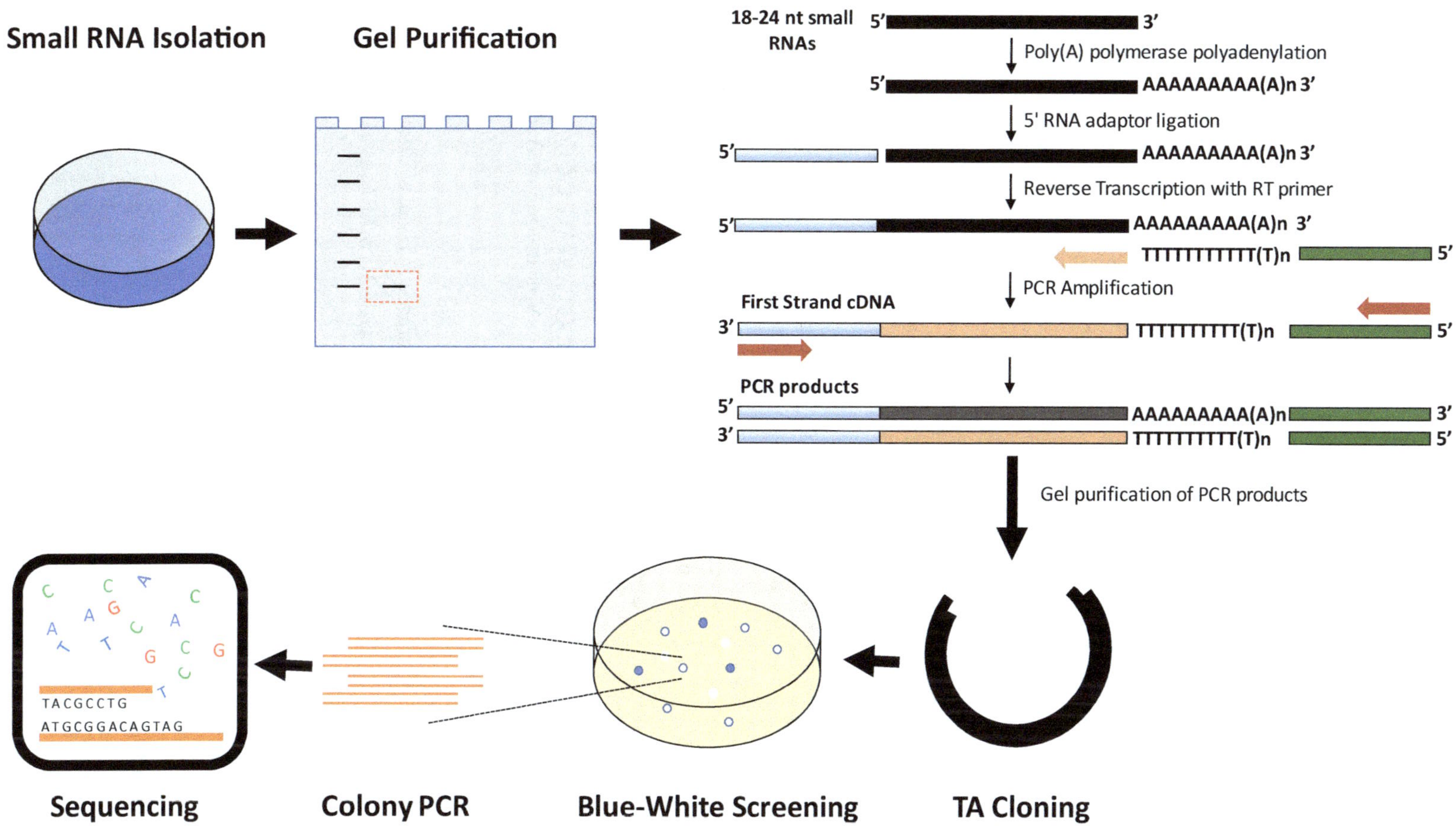

Fig. 2 Schematic depiction of the miRNA cloning protocol

1.8 M Tris base.
1.8 M Boric acid.
40 mM EDTA (pH 8.0).

3. 29:1 acryl:bis-acryl (30%) (Sigma).
4. Ammonium Persulfate (10% (w/v) in water).
5. *N,N,N′,N′*-tetramethylethane-1,2-diamine (TEMED) (Sigma).

2.5 Polyadenylation

1. *E.coli* Poly(A) Polymerase (NEB) (5000 U/ml).
2. 10× Polyadenylation Buffer.
3. ATP (10 mM) (store at −80 °C).
4. RiboLock RNAse Inhibitor (20 U/μL) (Thermo).

2.6 5′ Adapter Ligation

1. T4 RNA Ligase 1 (ssRNA Ligase) (NEB).
2. 10× T4 RNA Ligase 1 Buffer.
3. ATP (10 mM) (store at −80 °C).
4. RiboLock RNAse Inhibitor (20 U/μl) (Thermo).
5. 5′ RNA adapter (5′-CGA CUG GAG CAC GAG GAC ACU GAC AUG GAC UGA AGG AGU AGA AA-3′) (store at −80 °C).

2.7 Reverse Transcription

1. RevertAid Reverse Transcriptase (200 U/μl) (Thermo).
2. 5× RT Buffer (with 20 mM $MgCl_2$ and 50 mM DTT).
3. RiboLock RNAse Inhibitor (20 U/μl).
4. dNTP mix (10 mM).
5. (dT)30Rtprimer (5′-ATT CTA GAG GCC GAG GCG GCC GAC ATG-d(T)30VN-3′). V is anyone of A, G, or C, and N is anyone of A, G, C, or T.
6. RNAse H (Invitrogen).

2.8 PCR Amplification

1. Dream Taq DNA Polymerase (5 U/μl) (Thermo).
2. 10× Dream Taq Buffer (with 20 mM $MgCl_2$).
3. dNTP mix (10 mM).
4. 5′ cDNA primer (5′-GGA CAC TGA CAT GGA CTG AAG GAG TA-3′).
   3′ cDNA primer (5′-ATT CTA GAG GCC GAG GCG GCC GAC ATG T-3′).

2.9 TA Cloning and Colony PCR

1. InstaClone PCR Cloning Kit (Thermo).
2. T4 DNA Ligase (5 U/μl) (Thermo).
3. XL1-Blue Competent Cells.
4. M13 primer set.

Forward: 5′-CCCAGTCAk2CGACGTTGTAAAACG-3′.
Reverse: 5′-AGCGGATAACAATTTCACACAGG-3′.

5. 6× DNA Loading Dye (Thermo).

3 Methods

3.1 Cell Culture

1. Prepare and prewarm EBSS buffer to investigate the autophagic responses of miRNA expression of cultured cells under starvation conditions.
2. Discard cell media and wash cells twice in PBS.
3. For starvation, incubate cells in EBSS. As a control for nutrient-rich conditions, incubate cells in appropriate culture medium.
4. For pharmocological induction of autophagy, use Torin-1 is as mTOR complex inhibitor. Incubate cells in media containing Torin-1 (250 nM) that is dissolved in DMSO. For the controls, add the solvent DMSO to the medium.

3.2 Small RNA Isolation

We recommend the miRVana miRNA Isolation Kit for isolation of RNA <200 nt.

1. Aspirate the media and wash cells once with ice cold PBS. Remove PBS as much as possible.
2. Perform all the following steps according to the manual of the manufacturer.

3.3 Gel Purification of miRNA

To recover 18–24 nucleotide fraction of isolated small RNAs, perform denaturing UREA polyacrylamide gel electropheresis.

% Acrylamide	Seperation range (nt)	Bromophenol blue migration (nt)	Xylene cyanol migration (nt)
5	55–350	35	140
8	30–200	26	75
10	20–100	12	55
15	10–50	10	40
20	5–30	8	28

Following instructions are for a 1 mm wide, 7 cm mini-gel.

1. Pour 7 M urea, 15% PAGE gel according to the following protocol:

   Urea (7 M): 4.2 g.
   20× TBE: 500 μl.
   30% *Bis*-acrylamide (29:1): 4.97 ml.
   Add water to up to 10 ml.

Heat gently and swirl (use a 55 °C water bath) until urea dissolves. (*See* Note 2).
Add to 10 ml PAGE mix:
50 µl 10% APS, invert to mix.
10 µl TEMED, invert to mix.

2. Clean glass plates with water then ethanol.
3. Pour gel immediately and insert comb.
4. Wait for 30 min for polymerization of the gel.
5. After polymerization, warm the gel to ~50 °C by prerunning it for 20 min in 1× TBE running buffer. (*See* Note 3).
6. Carefully rinse the wells of the gel with 1× TBE. (*See* Note 4).
7. Mix isolated RNA sample with an equal volume of Gel Loading Buffer II and heat for 5 min at 95 °C to denature RNA and place in ice immediately.
8. Load the samples as well as 10 bp DNA ladder. Run the gel at 200 V for approximately 2 h until the bromophenol blue migrates about 4–5 cm to the gel. (*See* Note 5).
9. For visualization of the small RNAs, stain the gel with EtBr (10 mg/ml) in 1× TBE for 10 min at room temperature and visualize with UV transilluminator. Avoid excess UV exposure which will damage the nucleic acids.
10. Locate the band around 20 bp (which indicates miRNA), cut the band and smash into small pieces with sterile razor blade. (*See* Note 6).
11. Place smashed gel pieces in 1.5 ml eppendorf and add 500 µl of gel elution buffer.
12. Elute at room temperature for overnight to allow the RNA diffuse from gel.
13. Centrifuge the tube at maximum speed for 15 min at room temperature.
14. Place the supernatant into fresh tube and add 500 µl phenol–chloroform–isoamyl alcohol (25:24:1), centrifuge at 12000 × *g* for 10 min.
15. Transfer aqueous phase to a new tube and add 0.1 vol sodium acetate (3 M), 1 vol isopropanol or 2.5 vol 100% ethanol and 2 µl glycogen.
16. Precipitate RNA at −20 °C for overnight or −80 °C for 2 h.
17. After incubation, spin at 12000 × *g* for 30 min for precipitation and remove supernatant.
18. Wash the pellet with 750 µl 75% EtOH in DEPC, spin at 10000 × *g* at 4 °C for 15 min.
19. Remove the supernatant avoiding touching pellet. After air-drying, dissolve the pellet in 50 µl DEPC-H_2O.

3.4 Polyadenylation

Polyadenylation is required for preparing a priming site for cDNA synthesis using poly-dT on its 3′-end site.

Perform polyadenylation of small RNAs by *E.coli* Poly-A Polymerase.

Component	Volume (μl)
RNA	36
10× polymerase Rxn buffer	5
ATP (10 mM)	5
E.coli poly(A) polymerase	2
RiboLock RNAse inhibitor	2
Total	50

1. Incubate the reaction mix at 37 °C for 30 min.
2. Add 50 μl phenol–chloroform–isoamyl alcohol (25:24:1), centrifuge at 12000 × *g* for 10 min.
3. Transfer aqueous phase to a new tube and add 0.1 vol sodium acetate (3 M), 1 vol isopropanol or 2.5 vol 100% ethanol and 2 μl glycogen.
4. Precipitate polyadenylated RNA at −20 °C for overnight or −80 °C for 2 h.
5. After incubation, spin at 12000 × *g* for 30 min for precipitation and remove supernatant.
6. Wash the pellet with 750 μl 75% EtOH in DEPC, spin at 10000 × *g* at 4 °C for 15 min.
7. Remove the supernatant avoiding touching pellet. After air-drying, dissolve the pellet in 10 μl DEPC-H_2O.

3.5 5′ Adapter Ligation

5′ Adapter ligation is required as the priming site for amplification of first strand cDNA.

1. Ligate 5′ RNA adapter (5′-CGA CUG GAG CAC GAG GAC ACU GAC AUG GAC UGA AGG AGU AGA AA-3′) to the precipitated RNA with T4 RNA Ligase 1 (ssRNA Ligase). (*See* Note 7).

Component	Volume (μl)
RNA	10
10× T4 RNA ligase buffer	2
ATP (10 mM)	2
5′ RNA adapter (100 μM)	4
T4 RNA ligase	2
RiboLock RNAse inhibitor	1
Total	20

2. Incubate the reaction mix at room temperature for 3 h or at 16 °C for overnight in a thermal cycler. (*See* Note 8).
3. Add 20 μl phenol–chloroform–isoamyl alcohol (25:24:1), centrifuge at 12000 × *g* for 10 min.
4. Add 0.1 vol sodium acetate (3 M), 1 vol isopropanol or 2.5 M vol 100% ethanol and 2 μl glycogen.
5. Precipitate polyadenylated and 5′ adapter ligated RNA at −20 °C for overnight or −80 °C for 2 h.
6. After incubation, spin at 12000 × *g* for 30 min for precipitation and remove supernatant.
7. Wash the pellet with 750 μl 75% EtOH in DEPC, spin at 10000 × *g* at 4 °C for 15 min.
8. Remove the supernatant avoiding touching pellet. After air-drying, dissolve the pellet in 10 μl DEPC-H_2O.

3.6 Reverse Transcription

Convert single-stranded RNA to single-stranded cDNA using RevertAid First Strand cDNA Synthesis Kit and small RNA d(T) 30 RT-Primer (5′-ATT CTA GAG GCC GAG GCG GCC GAC ATG-d(T)30VN-3′).

Add the following components into a sterile, nuclease- free tube on ice in the indicated order:

Component	Volume (μl)
Polyadenylated and 5′ adaptor ligated RNA	10
d(T)30 RT primer (100 μM)	0.5
dNTP (10 mM)	1
Total	11.5

1. Incubate the tube at 65 °C for 5 min and chill on ice (*See* Note 9).
2. Add the following components on to the previous tube in the indicated order:

Component	Volume (μl)
5× RT buffer	4
RevertAid	2
RiboLock RNAse inhibitor	1
Nuclease-free water	1.5
Total	20

3. Incubate the tube at 42 °C for 60 min.
4. Terminate the reaction by heating at 70 °C for 5 min.
5. Add 1 μl of RNAse H and incubate at 37 °C for 20 min to degrade RNA in the reaction mixture.

3.7 PCR Amplification

Add the following components into a sterile, thin walled PCR tube on ice in the indicated order:

Component	Volume (μl)
cDNA	10
10× DreamTaq buffer	5
dNTP (10 mM)	1
5′ cDNA primer (50 μM)	1
3′ cDNA primer (50 μM)	1
DreamTaq	0.5
Nuclease-free water	To 50
Total	50

1. Place the tubes in a thermal cycler. Perform PCR performed according to the following thermal cycling conditions:
   (a) 95 °C 3 min.
   (b) 95 °C 30 s, 50 °C 30 s, 72 °C 30 s, 30 cycles.
   (c) 72 °C 5 min.
   (d) 4 °C infinity.

3.8 Gel Analysis of Amplified cDNAs

To separate PCR products, perform nondenaturing TBE–acrylamide gel electrophoresis.

% Acrylamide	Seperation range (bp)	Bromophenol blue migration (bp)	Xylene cyanol migration (bp)
5	80–500	65	260
8	60–400	45	160
12	40–200	20	70
15	25–100	15	60
20	5–100	12	45

Following instructions are for a 1 mm wide, 7 cm mini-gel.

1. Pour 12% nondenaturing TBE–acrylamide gel according to the following protocol:

   20× TBE: 600 μl.
   30% *Bis*-acrylamide (29:1): 4.8 ml.
   10% APS: 200 μl.
   TEMED: 10 μl.
   Water: 6.4 ml.
   Total: 12 ml.

2. Pour gel immediately and insert comb. Wait for 30 min for polymerization of the gel.
3. Rinse the wells of the gel with 1× TBE.
4. Mix 15 μl of PCR products with 3 μl 6× DNA Loading Dye.
5. Load the samples as well as 100 bp DNA ladder. Run the gel at 200 V for approximately 2 h until the bromophenol blue migrates at the end of the gel.
6. For visualization of the PCR products, stain the gel with EtBr (10 mg/ml) in 1× TBE for 30 min at room temperature and visualize with UV transilluminator.
7. Calculate the size based on the adapter and PCR primers. In this case, around 100 bp band indicates amplified cDNAs from miRNA. Cut the band corresponding to that size using a clean scalpel.
8. Place smashed gel pieces in 1.5 ml eppendorf and add 500 μl of gel elution buffer pieces.
9. Elute at room temperature for overnight.
10. Centrifuge the tube at maximum speed for 15 min at room temperature.
11. Place the supernatant into fresh tube and add 500 μl phenol–chloroform–isoamyl alcohol (25:24:1), centrifuge at 12000 × *g* for 10 min.

12. Transfer aqueous phase to a new tube and add 0.1 vol sodium acetate (3 M), 1 vol isopropanol or 2.5 vol 100% ethanol and 2 μl glycogen.
13. Precipitate amplified cDNA at −20 °C for overnight or −80 °C for 2 h.
14. After incubation, spin at 12000 × *g* for 20 min for precipitation and remove supernatant.
15. Wash the pellet with 750 μl 75% EtOH, spin at 10000 × *g* at 4 °C for 15 min.
16. Remove the supernatant avoiding touching pellet. After air-drying, dissolve the pellet in 30 μl TE buffer.

3.9 TA Cloning and Colony PCR

In order to generate high-quality miRNA library, it is crucial to perform efficient cloning. Perform cloning with InsTAclone PCR Cloning Kit using XL1-Blue competent cells with high transformation efficiency.

Prepare ligation mixture according to the following:

Component	Volume (μl)
Vector pTZ57R/T	3
5× ligation buffer	6
PCR product	5
T4 DNA ligase	1
Nuclease-free water	To 30
Total	30

1. Incubate the ligation mixture at 16 °C overnight.
2. Combine 10 μl of ligation mixture with XL1-Blue competent cells for bacterial transformation.
3. Prepare X-gal/IPTG LB Agar plates blue/white colony screening. Add screening antibiotic of choice.
4. Spread 50 μl of transformation mix and incubate at 37 °C overnight. (*See* Note 10).
5. Pick white colonies as colonies with insert into 20 μl PCR mix.

Component	Volume (μl)
10× DreamTaq buffer	2
dNTP (10 mM)	1
M13 forward primer	0.6
M13 reverse primer	0.6
DreamTaq	0.2
Nuclease free water	To 20
Total	20

Make a short strike with the same tip over culture plate to save the clone for repropagation.

6. Perform colony PCR with M13 Forward and Reverse primers according to the following thermal cycling conditions:
   (a) 95 °C 3 min.
   (b) 95 °C 30 s, 50 °C 30 s, 72 °C 30 s, 30 cycles.
   (c) 72 °C 5 min.
   (d) 4 °C infinity.
7. Prepare 2% agarose gel; mix 10 μl of each sample with 2 μl of 6× Loading Dye. Load samples and 100-bp ladder on the gel.
8. Visualize the gel with UV transilluminator. Most colonies will generate a band near 250 bp.
9. Amplicons with insert are ready for sequencing with M13 primers. (*See* Note 11).

4 Notes

1. This buffer is available commercially but if homemade is needed, the below is the recipe for 2× RNA gel loading dye.

   95% Formamide.
   0.025% (w/v) SDS.
   0.025% Bromophenol Blue.
   0.025% Xylene Cyanol.
   5 mM EDTA pH 8.0.
   Prepare in DEPC-treated of nuclease-free water.
2. 7 M Urea ensures the linearity of the nucleic acids by avoiding secondary/tertiary structure formation. Cool down the gel by placing on ice for approximately 5 min before adding 10% APS and TEMED. Otherwise, polymerization of the gel will occur too quickly before you pour the gel. Extended cooling may result in precipitation of urea in a crystal form.

3. Choose constant watts at 15–25 W per gel for the prerun in order to remove remaining urea from the gel. The optimal temperature should be between 45 and 55 °C degrees.
4. Since urea will diffuse from the wells, create a layer at the bottom of the gel and prevent sample loading, it is important to rinse the wells carefully with 1× TBE solution.
5. Heating at 45–55 °C is necessary for denaturing samples during running. Excess heating over 65 °C should be avoided since it will disrupt the gel and gel apparatus.
6. High quality small RNA will have clearly visible tRNA, 5S rRNA, and 5.8S rRNA bands. We find that it is best to run 100 pmol of a miRNA oligo near with small RNA samples since it will locate the exact band of 18–24 nucleotides.
7. Ligation efficiency is often a problem for cloning. We find that it is best to perform T4 RNA ligation at 4–16 °C overnight that could significantly improve the ligation efficiency.
8. Denaturation of RNA template and primer in the absence of reaction buffer and enzyme is necessary to remove secondary structure that may prevent proper cDNA synthesis.
9. Control DNA 1 (vector without insert) can be used as negative control and control DNA 2 (the cloning vector with inserted control PCR fragment) can be used as a positive control.
10. If there exists sequence errors in the cloned insert, this might due to damage of PCR product by UV light during excision from agarose gel. We prefer to use a long wavelength UV (360 nm) light-box when excising DNA from the agarose gel. Dyes such as crystal violet which is less mutagenic but less sensitive than ethidium bromide might be used to visualize DNA in ambient light.

References

1. Ravikumar B, Sarkar S, Davies JE, Futter M, Garcia-Arencibia M, Green-Thompson ZW et al (2010) Regulation of mammalian autophagy in physiology and pathophysiology. Physiol Rev 90(4):1383–1435. https://doi.org/10.1152/physrev.00030.2009
2. Gozuacik D, Kimchi A (2004) Autophagy as a cell death and tumor suppressor mechanism. Oncogene 23(16):2891–2906. https://doi.org/10.1038/sj.onc.1207521
3. Pan H, Cai N, Li M, Liu G, Belmonte J (2013) Autophagic control of cell stemness'. EMBO Mol Med 5(3):327–331. https://doi.org/10.1002/emmm.201201999
4. Liu F, Lee JY, Wei H, Tanabe O, Engel JD, Morrison SJ et al (2010) FIP200 is required for the cell-autonomous maintenance of fetal hematopoietic stem cells. Blood 116:4806–4814. https://doi.org/10.1182/blood-2010-06-288589
5. Salemi S, Yousefi S, Constantinescu MA, Fey MF, Simon H-U (2012) Autophagy is required for self-renewal and differentiation of adult human stem cells. Cell Res 22:432–435. https://doi.org/10.1038/cr.2011.200
6. Vazquez P, Arroba AI, Cecconi F, de la Rosa EJ, Boya P, de Pablo F (2012) Atg5 and Ambra1 differentially modulate neurogenesis in neural stem cells. Autophagy 8:187–199. https://doi.org/10.4161/auto.8.2.18535
7. Tra T, Gong L, Kao LP, Li XL, Grandela C, Devenish RJ et al (2011) Autophagy in human

embryonic stem cells. PLoS One 6:e27485. https://doi.org/10.1371/journal.pone.0027485

8. Menendez JA, Vellon L, Oliveras-Ferraros C, Cufi S, Vazquez-Martin A (2011) mTOR-regulated senescence and autophagy during reprogramming of somatic cells to pluripotency: a roadmap from energy metabolism to stem cell renewal and aging. Cell Cycle 10:3658–3677. https://doi.org/10.4161/cc.10.21.18128
9. Kim VN (2005) MicroRNA biogenesis: coordinated cropping and dicing. Nat Rev Mol Cell Biol 6(5):376–385. https://doi.org/10.1038/nrm1644
10. Gozuacik D, Akkoc Y, Ozturk DG, Kocak M (2017) Autophagy-regulating microRNAs and cancer. Front Oncol 7. https://doi.org/10.3389/fonc.2017.00065
11. Barad O, Meiri E, Avniel A, Aharonov R, Barzilai A, Bentwich I et al (2004) MicroRNA expression detected by oligonucleotide microarrays: system establishment and expression profiling in human tissues. Genome Res 14(12):2486–2494
12. Chen C, Ridzon DA, Broomer AJ, Zhou Z, Lee DH, Nguyen JT et al (2005) Real-time quantification of microRNAs by stem-loop RT-PCR. Nucleic Acids Res 33(20):e179
13. Lee RC, Ambros V (2001) An extensive class of small RNAs in Caenorhabditiselegans. Science 294(5543):862–864
14. Lau NC, Lim LP, Weinstein EG, Bartel DP (2001) An abundant class of tinyRNAs with probable regulatory roles in Caenorhabditis elegans. Science 294(5543):858–862
15. Lagos-Quintana M, Rauhut R, Yalcin A, Meyer J, Lendeckel W, Tuschl T (2002) Identification of tissue-specific microRNAs from mouse. Curr Biol 12(9):735–739
16. Lu C, Meyers BC, Green PJ (2007) Construction of small RNA cDNA libraries for deep sequencing. Methods 43:110–117. https://doi.org/10.1016/j.ymeth.2007.05.002
17. Gu W, Shirayama M, Conte D, Vasale J, Batista PJ, Claycomb JM et al (2009) Distinct argonaute-mediated 22G-RNA pathways direct genome surveillance in the C. elegans germline. Mol Cell 36:231–244. https://doi.org/10.1016/j.molcel.2009.09.020
18. Fu H, Tie Y, Xu C, Zhang Z, Zhu J, Shi Y, Jiang H, Sun Z, Zheng X (2005) Identification of human fetal liver miRNAs by a novel method. FEBS Lett 579(17):3849–3854. https://doi.org/10.1016/j.febslet.2005.05.064

Methods in Molecular Biology (2018) 1854: 147–157
DOI 10.1007/7651_2018_157

Published online: 5 June 2018

Assays to Monitor Aggrephagy in *Drosophila* Brain

Anne-Claire Jacomin and Ioannis P. Nezis

Abstract

Accumulation of ubiquitinated protein aggregates is a hallmark of most aging-related neurodegenerative disorders. Autophagy has been found to be involved in the selective clearance of these protein aggregates, and this process is called aggrephagy. Here we provide two protocols for the investigation of protein aggregation and their removal by autophagy using western blotting and immunofluorescence techniques in *Drosophila* brain. Investigating the role of aggrephagy at the cellular and organismal level is important for the development of therapeutic interventions against aging-related diseases.

Keywords Aggregate, Aggrephagy, Brain, *Drosophila*, Immunofluorescence, Protein extraction, Ref(2)P, Ubiquitin, Western blot

1 Introduction

Autophagy is a dynamic catabolic pathway based on the formation of double-membraned vesicles (autophagosomes) that eventually fuse with the lysosomes for the degradation of their content [1]. This pathway has been related to many physiological processes that are involved in the maintenance of cellular homeostasis and quality control of the cell by the removal of redundant or damaged proteins and organelles [2].

Damaged proteins often form large aggregates that are ubiquitinated. Selective removal of these aggregates by autophagy is called aggrephagy [3]. Ubiquitination may facilitate the uptake of aggregates into autophagosomes. Indeed, aggrephagy receptors such as p62/SQSTM1, NBR1, NDP52/CALCOCO2, optineurin, and Tollip recognize ubiquitin via their ubiquitin-binding domains [3–6].

The most described selective autophagy receptor in *Drosophila* is the protein Ref(2)P is homolog to mammalian p62/SQSTM1. Like its mammalian counterpart, Ref(2)P contains multiple interaction domains, including a ubiquitin-binding domain UBA and a LIR motif crucial for its connection to autophagy through the interaction with Atg8a (homolog to LC3/GABARAP mammalian proteins) [7, 8].

Autophagy is essential for long-lived cells such as neurons. Indeed, neurons are post-mitotic cells and cannot dilute their cytosol by cell division, thus relying on autophagy for the elimination of aggregates. Dysfunctional autophagy has been associated with premature aging and neurodegeneration [9]. In that aspect, *Drosophila* constitutes a model of choice due to its short life cycle, rapid aging, and easy and low-cost maintenance.

In this chapter, we describe two protocols for the investigation of aggregate formation in adult *Drosophila* brain. The first technique consists in the global analysis of protein aggregation by differential detergent protein extraction and Western blotting. Indeed, unlike their non-aggregation-prone counterpart, aggregated proteins are often characterized by being insoluble in lysis buffers containing mild detergents. Also, it is possible to investigate the turnover of aggregated protein by extraction of soluble proteins using a mild detergent-containing lysis buffer, followed by solubilizing the pelleted material under harsher conditions with a lysis buffer containing 2% SDS [10].

The second protocol described the dissection and immunostaining of *Drosophila* brain. Fluorescence microscopy is another widely used technique for the study of aggregated proteins. The advantage in fluorescence microscopy resides in the in situ staining and observation of the aggregate proteins in tissues, organs, and cells.

2 Materials

2.1 Drosophila Stocks and Maintenance

Flies are raised on a yeast/cornmeal diet and are kept at 25 °C and 70% humidity with a 12–12-h light-dark cycle. The fly stocks used in this chapter are available from the Bloomington *Drosophila* Stock Center (Indiana University): w^{1118} (#3605, referred to as "WT") and $Atg8a^{KG07569}$ (#14639, referred to as "Atg8a").

Fly food: 1 L H_2O, 42 g inactive dry yeast, 60 g yellow cornmeal, 130 g sucrose, 5.5 g agar, and 15 mL Nipagin 10% (*see* **Note 1**).

2.2 Buffers and Solutions for Protein Extraction and Western Blotting

1. Triton lysis buffer: 1× PBS, 1% Triton X-100, 1 mM Na_3VO_4, 50 mM NaF, 5 mM $Na_4P_2O_7$, 1× protease inhibitor cocktail.
2. SDS lysis buffer: 50 mM Tris pH 7.8, 2% (w/v) SDS, 1 mM Na_3VO_4, 50 mM NaF, 5 mM $Na_4P_2O_7$, 1× protease inhibitor cocktail.
3. 10× Gel protein running buffer: 250 mM Tris base, 2 M glycine, 1% SDS, pH 8.3.
4. Transfer buffer: 25 mM Tris, 2 mM glycine, 10% ethanol. Prepare a 10× transfer buffer stock solution (250 mM Tris base, 2 M glycine, pH 8.3). Dilute one part of 10× stock

solution with eight parts of water and one part of ethanol (*see* **Notes 2** and **3**).

5. 6× Laemmli buffer stock solution: 12% (w/v) SDS, 60% glycerol, 0.12% (w/v) bromophenol blue, 0.375 M Tris pH 6.8 (*see* **Note 4**).
6. 4× Laemmli buffer working solution: 6.7 mL 6× Laemmli, 2 mL 2-mercaptoethanol (20%), 1.3 mL ddH_2O (*see* **Note 5**).
7. TBS (10×, 1 L): 24.23 g Trizma HCl, 80.06 g NaCl, pH 7.6.
8. TBST: 1× TBS, 0.1% Tween-20.
9. Blocking buffer: 5% (w/v) nonfat milk or BSA in 1× TBST.
10. Secondary antibody incubation buffer: 1% (w/v) nonfat milk or BSA in 1× TBST.
11. Mild stripping buffer (1 L): 15 g glycine, 1 g SDS, 10 mL Tween-20, pH 2.2.

2.3 Buffers and Solution for Immunofluorescence

1. Fixative: 4% methanol-free paraformaldehyde in 1× PBS (*see* **Note 6**).
2. Phosphate-buffered saline (PBS) (1× PBS pH 7.4).
3. PBS-T (wash buffer): 1× PBS, 0.1% Tween-20.
4. Permeabilization buffer: 1× PBS, 0.1% Triton X-100, 0.3% BSA (*see* **Note 7**).
5. Hoechst dissolved in distilled water to obtain a 1 mM (1000×) stock solution, which can be stored in a dark vial at 4 °C.
6. Mounting medium: 70% glycerol, 2% w/v propyl gallate, 1× PBS. Can be stored for 6 months in a dark vial at 4 °C.

2.4 Specific Equipment

2.4.1 Protein Extraction and Western Blotting

1. Cryotubes.
2. Funnels, chilled at −80 °C.
3. Tea strainers with two different mesh size, chilled at −80 °C.
4. Pestle motor and pestles.
5. Sonicator.

2.4.2 Immunofluorescence

1. Fine tweezers and sharpening stone.
2. Deep-well glass slide.
3. 48-Well plate and Terasaki plate.
4. Basket made by cutting a cap from cell strainer tube.
5. Microscope slides and coverslips (22 × 22 mm).

3 Methods

3.1 Preparation of the Flies

1. Flies from the appropriate genotype are collected within 24 h from hatching and packed in cohorts of maximum 25 flies.
2. Flies are transferred into new tubes every 2–3 days until they reach the desired age (5–7 days old for young flies, 30–50 days old for old flies) (*see* **Note 8**).

3.2 Differential Detergent Protein Extraction Analysis (Fig. 1)

3.2.1 Differential Detergent Protein Extraction

1. Flash freeze the flies in liquid nitrogen (*see* **Notes 9** and **10**).
2. Decapitate the flies by vortexing the tubes twice (*see* **Note 11**).
3. Separate the heads from the bodies and other small body parts such as legs and antenna using two tea strainers with different mesh size, starting with the smaller size to remove the small parts. Using the larger mesh and a funnel, collect the heads in a new tube.
4. Lyse the heads in chilled Triton lysis buffer for protein extraction from the soluble fraction. Homogenize with a pestle until heads can no longer be discerned (about 20 s when using a motorized pestle) (*see* **Note 12**).
5. Centrifuge lysate at 17,000 × *g* for 15 min at 4 °C in a prechilled centrifuge.
6. Save the supernatant; this fraction contains the soluble proteins (Fig. 1a).
7. Wash the pellet once with 200 μL of Triton lysis buffer.
8. Centrifuge at 17,000 × *g* for 2 min at 4 °C in a prechilled centrifuge, and discard the supernatant.
9. Add 100 μL of SDS lysis buffer onto the pellet for protein extraction from the insoluble fraction (*see* **Note 13**).
10. Sonicate on ice until the pellet appears to be lysed.
11. Centrifuge lysate at 13,000 × *g* for 10 min.
12. Save the supernatant; this fraction contains the insoluble proteins originally contain within aggregates (Fig. 1a).
13. Quantitate the amount of protein using a kit or assay of choice.
14. Prepare the loading samples in 2× or 4× Laemmli buffer as for convenience.

3.2.2 Gel Electrophoresis and Western Blotting

1. Separate 10–20 μg of protein by SDS-PAGE at constant voltage per gel (*see* **Notes 14–17**).
2. Transfer proteins onto nitrocellulose or PVDF membranes in a prechilled transfer buffer that contains 10% ethanol (*see* **Notes 18–20**).
3. Block the membrane in blocking buffer for 1 h at room temperature with gentle rocking to reduce nonspecific antibody binding.

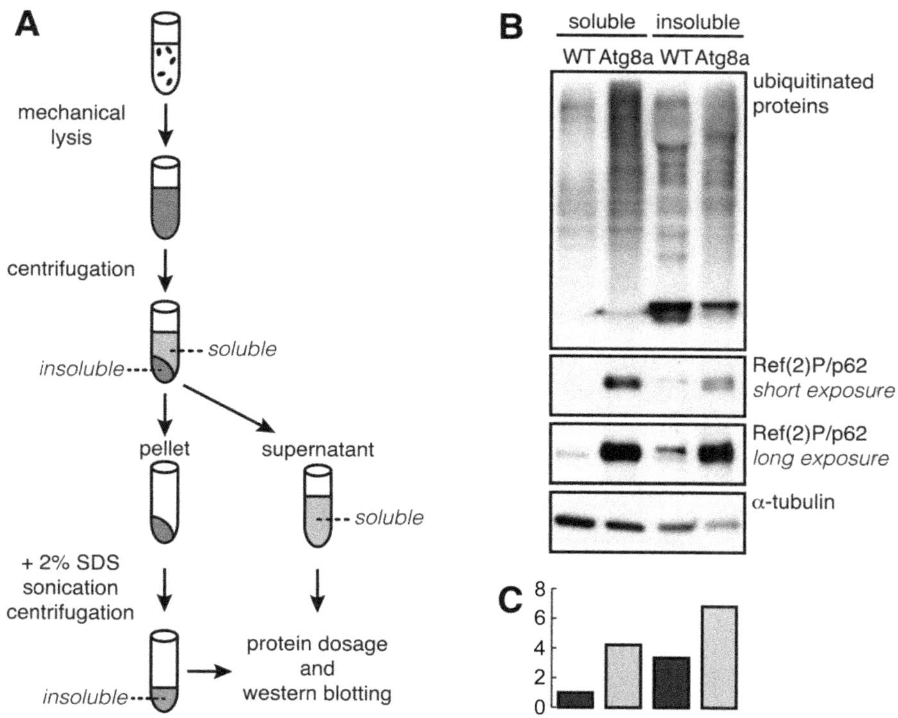

Fig. 1 Differential detergent protein extraction. (**a**) Flowchart of the procedure described in Subheading 3. (**b**) Differentially extracted proteins from young wild-type (WT) and autophagy-deficient fly heads (Atg8a) were subjected to immunoblotting using anti-ubiquitinated proteins and Ref(2)P antibodies. Tubulin was used as a loading control. (**c**) Quantification of Ref(2)P relative protein level by densitometry using Fiji/ImageJ as described in Subheading 3

4. Incubate in primary antibody in TBST overnight at 4 °C with gentle rocking (*see* **Notes 21** and **22**).
5. Wash the membrane four times (5 min each) in 1× TBST with gentle rocking at room temperature.
6. Dilute horseradish peroxidase-conjugated secondary antibody in TBST + 1% nonfat milk or BSA. Incubate with secondary antibody for 45 min at room temperature with gentle rocking (*see* **Note 23**).
7. Wash the membrane three times (5–10 min each) with TBST with gentle rocking.
8. Rinse once with TBS (*see* **Note 24**).
9. Remove the excess of buffer, and cover the membrane with chemiluminescence reagent, in accordance with manufacturer's recommendations.
10. Document the western blotting results (X-ray film or digital imaging) (Fig. 1b).

3.2.3 Membrane Stripping and Reprobing

1. After imaging, place the membrane back into TBS and wash twice in TBS to remove the chemiluminescence reagent.
2. Cover the membrane with mild stripping buffer and incubate at room temperature for 5 min with gentle agitation.
3. Discard the stripping buffer, and repeat incubation with fresh mild stripping buffer for 10 min.
4. Wash twice the membrane in TBS (at least 10 min each) to ensure all the stripping buffer is washed away.
5. Wash twice the membrane in TBST (5–10 min each).

The membrane is now ready for blocking and immunostaining as described in Subheading 3.2.2.

3.2.4 Analysis Using ImageJ/Fiji

Several methods exist for protein quantitation by densitometric analysis using ImageJ/Fiji, which give similar results. One of these methods, using the "Gel analysis" tool embedded in the software, is detailed hereafter.

Before proceeding to densitometric analysis, the X-ray films must be scanned in grayscale mode at a DPI of 300 or greater and saved in TIFF format.

1. Open the scan file, and using the LUT option in the main toolbar, make sure that none of the band to quantify is over-saturated (*see* **Note 25**).
2. If necessary, the background noise can be reduced; go to the menu under Process>Subtract Background, and start with a rolling ball radius of 50. This may need to be adjusted for each densitometric analysis.
3. On the toolbar, select the rectangle selection tool, and draw a rectangle around the first band. From the Analyze>Gels submenu, choose the "Select First Lane" command, or type the "1" key on the keyboard, to designate the first lane. Move the rectangle to the next lane, and go to Analyze>Gels>Select Next Lane or press the "2" key, and continue to do this for each subsequent lane.
4. Once all the lanes are highlighted, select the command "plot lanes" from the Analyze>Gels submenu. This will open a new window displaying the densitometry histogram for each lane.
5. Use the line tool to draw two vertical lines, one on each side of the peak, to enclose the peak area.
6. Select the magic wand from the toolbar and click on each peak. This will open a new result window gathering the area for each selected peak. This data can be copy and paste into an Excel spreadsheet.
7. Proceed the same way for all the proteins you need to quantify as well as the loading control.

8. In Excel, calculate the ratio between the peak area of a protein of interest and its corresponding loading control. For ease of interpretation, normalize the ratio to a control condition to get a "fold increased protein quantity" for the test condition (Fig. 1c).

3.3 Immunofluorescence of Aggregates in Adult Drosophila Brain Using Ref(2)P and Ubiquitin Staining (Fig. 2)

3.3.1 Adult Brain Dissection

1. Moisten a piece of tissue with water. This will be used to wipe the tweezers and remove debris during the dissection (*see* **Note 26**).
2. Anesthetize the flies on a CO_2 pad. Transfer a fly into a drop of PBS in a deep well, and orient it ventral side up. With one of the tweezer, grab the thorax just below the head. Using the second tweezer, grab the extended proboscis to remove the head. Discard the body and submerge the head (*see* **Notes 27** and **28**) (Fig. 2A(a, b)).
3. With the free forceps, obtain a grip between the proboscis and the eye. Alternating left and right, gently make use of both of your forceps to tear apart the retina, working toward the back of the head to remove the cuticle.
4. After isolating the brain, carefully remove the trachea (*see* **Note 29**).

3.3.2 Immunostaining and Mounting

1. Transfer the isolated brains into a basket, and immerse it in a well of the 48-well plate containing the fixative (Fig. 2A(c, e)). Incubate for 30 min at room temperature.
2. Rinse three times with 1× PBS (5 min each) by transferring the basket from one well to another.
3. Block in permeabilization buffer for 1 h at room temperature.
4. Carefully transfer the brains into a well of a Terasaki plate filled with 10–15 μL of antibody solution in the permeabilization buffer. Incubate overnight at 4 °C with a piece of tissue soaked in water (*see* **Note 30**) (Fig. 2A(d)).
5. Transfer back the brains into the basket and wash three times in PBS-T at room temperature with gentle agitation (10 min each).
6. Dilute the secondary antibody in the permeabilization buffer and incubate for 2 h at room temperature (*see* **Notes 31** and **32**).
7. Wash three times in PBS-T at room temperature with gentle agitation (10 min each).
8. Transfer the basket into a well containing the Hoechst solution (1× dilution in PBS) and incubate for 15 min at room temperature with gentle agitation.
9. Rinse once in PBS.

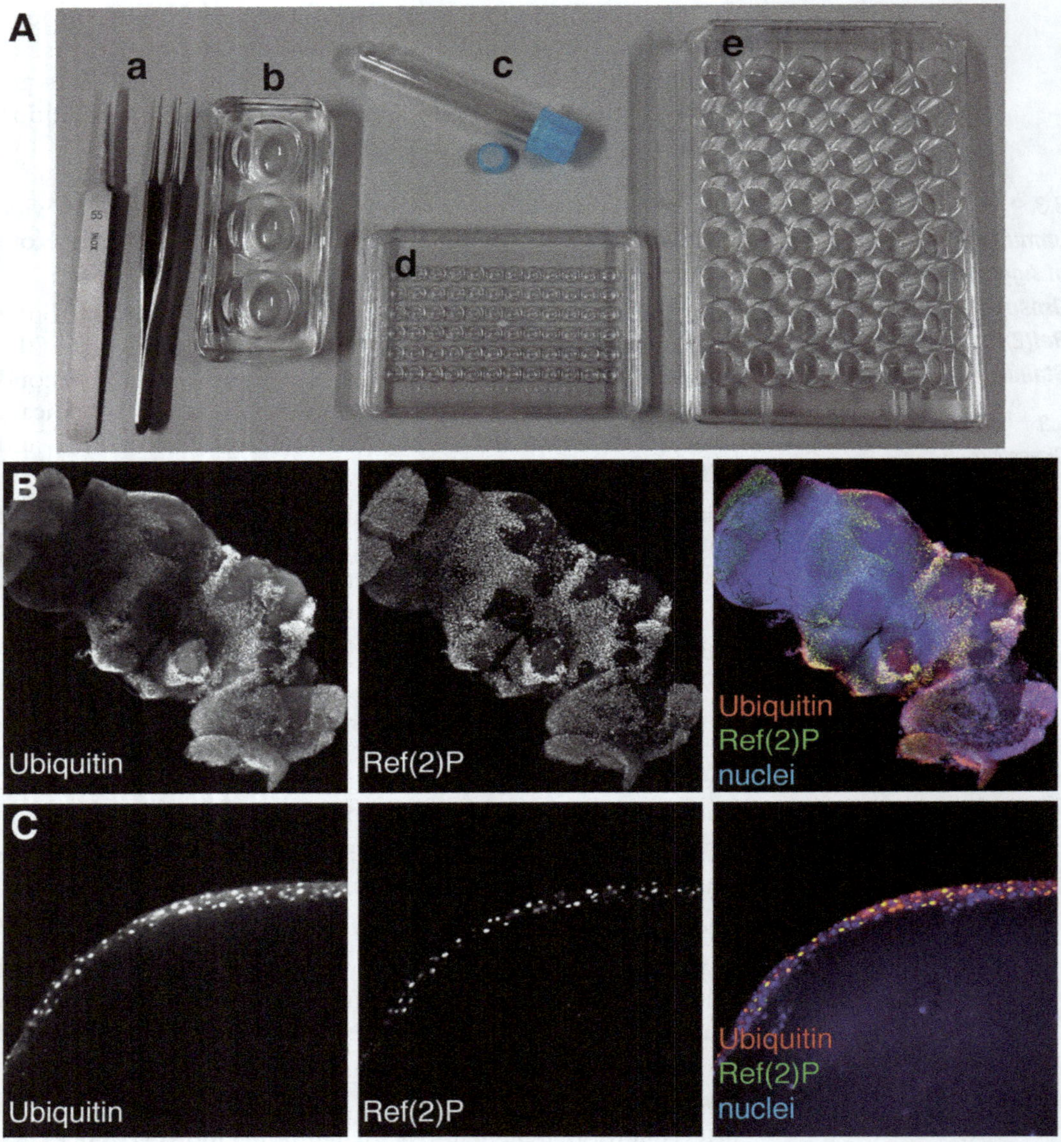

Fig. 2 Double staining of aggregates in adult brains from autophagy-deficient flies. (**A**) Tools for dissection and staining: fine tweezers (a), deep-well glass slide (b), small baskets created by cutting in the cap of a cell strainer tube (c), Terasaki plate, (d) and 48-well plate (e). (**B**, **C**) Confocal micrographs of *Drosophila* autophagy Atg8a mutant adult brain showing the presence of protein aggregates due to defective autophagy. (**B**) Low magnification of the brain and (**C**) a zoomed area of the optic lobe (**C**) after staining for ubiquitinated proteins (red), Ref(2)P (green), and nuclei (blue)

10. On a microscope slide, put a drop of mounting medium (*see* **Note 33**).
11. Carefully transfer the brains onto the slide, and orient them with their antennal lobes facing upward.

12. Gently cover with a coverslip to avoid the formation of bubbles.
13. Once the mounting medium has completely spread beneath the coverslip, seal the edges with nail polish.
14. Store the slides in a dark box at 4 °C.
15. Visualize using a confocal microscope. We preconize using first a low magnification lens (10×) to get global micrographs of the brains before moving to higher magnification lenses (typically 63×) (Fig. 2B, C). As much as possible, view multiple confocal sections within a tissue. When comparing samples, examine similar regions within the brains.

4 Notes

1. The quantity of agar needs to be adjusted in-house depending on the provider.
2. Ethanol proportion may vary from 5% (large proteins) to 20% (small proteins) depending on the size of the main protein of interest.
3. We do not use methanol for our transfer but prefer ethanol as it is less toxic and can be put down the sink safely.
4. Because of the high concentration of SDS, the 6× Laemmli buffer can become viscous due to the precipitation of the SDS. The buffer can be warmed up in a water bath at 60 °C before use.
5. For ease of use, 4× Laemmli buffer with 2-mercaptoethanol can be aliquoted and stored at −20 °C for a few months.
6. The 4% fixative solution can be stored at −20 °C for a few months.
7. The concentration of Triton X-100 can vary from 0.1% to 0.3% and needs to be optimized for each antibody.
8. Flies with a defective autophagy have a much shorter life span than their isogenic wild-type counterpart. So, for working with old autophagy-deficient flies, both wild-type and autophagy-deficient flies are frozen at 30 days from hatching.
9. Use SafeLock 1.5 mL Eppendorf tubes or cryotubes for freezing in liquid nitrogen safely.
10. If the flies are not processed immediately, store at −80 °C.
11. Vortex for a few seconds, and keep the tubes on dry ice between rounds of vortexing to avoid thawing the flies.
12. The number of head determines the amount of lysis buffer to add. We suggest starting with 20–30 heads in 200 μL of lysis buffer.
13. Add the SDS right before use as it will precipitate if kept on ice.

14. To get a good separation of Ref(2)P and tubulin, we use 8% acrylamide gels.
15. We usually run the SDS-PAGE at 50–60 V until the proteins cross the stacking gel and then increase to the desired voltage (up to 200 V).
16. When probing for ubiquitinated protein, we recommend running the SDS-PAGE at low voltage (max 100 V).
17. SDS-PAGE are run at a constant voltage to minimize the heating of the gels' chamber.
18. The usual transfer is done at 100 V for 1 h, except for ubiquitinated proteins: 25 V for 2 h.
19. When transferring onto PVDF membrane, it is crucial to incubate them for 5–10 min in 100% ethanol before assembling the transfer sandwich.
20. The use of a constant voltage ensures that field strength remains constant, thus providing a most efficient wet transfer.
21. The incubation time for the primary antibody may vary and must be optimized depending on the antibody and protein concentration.
22. Primary antibody buffer can contain 1% nonfat milk or BSA to reduce the nonspecific binding. Note that some antibodies behave better in BSA than milk.
23. We do not recommend to incubate the secondary antibody for more than 1 h to avoid nonspecific binding.
24. It is important to rinse the membranes once in TBS without Tween-20 as it can interfere with the chemiluminescence reagent.
25. It is crucial that the bands are not oversaturated as it will result in an underestimation of the densitometry.
26. Sharp forceps are essential for live dissections. Using a sharpening block, gently pass the forceps back and forth on each side until they are sharp enough.
27. It is imperative to always hold the head firmly with tweezers; otherwise, it will float and is difficult to retrieve.
28. If the head moves out of focus, move it back into the focal plane without adjusting the microscope.
29. It is crucial to remove as much as possible the trachea as those air-filled sacs can cause brains to float and affect the staining and imaging.
30. Adding a piece of tissue soaked in water will contribute to reducing the evaporation of the antibody solution.
31. The incubation with the secondary antibodies can also be conducted overnight at 4 °C.

32. From that point, it is crucial to perform all the incubation in a dark box to avoid bleaching of the staining.
33. For a coverslip 22 × 22 mm, a 15 μL drop of mounting medium is sufficient.

Acknowledgments

This work is supported by BBSRC grants BB/L006324/1 and BB/P007856/1 awarded to Dr. Ioannis P. Nezis.

References

1. Feng Y, He D, Yao Z, Klionsky DJ (2014) The machinery of macroautophagy. Cell Res 24 (1):24–41
2. Mizushima N, Komatsu M (2011) Autophagy: renovation of cells and tissues. Cell 147 (4):728–741
3. Lamark T, Johansen T (2012) Aggrephagy: selective disposal of protein aggregates by macroautophagy. Int J Cell Biol 2012:736905
4. Kirkin V, Lamark T, Sou YS, Bjorkoy G, Nunn JL, Bruun JA, Shvets E, McEwan DG, Clausen TH, Wild P, Bilusic I, Theurillat JP, Overvatn A, Ishii T, Elazar Z, Komatsu M, Dikic I, Johansen T (2009) A role for NBR1 in autophagosomal degradation of ubiquitinated substrates. Mol Cell 33(4):505–516
5. Pankiv S, Clausen TH, Lamark T, Brech A, Bruun JA, Outzen H, Overvatn A, Bjorkoy G, Johansen T (2007) p62/SQSTM1 binds directly to Atg8/LC3 to facilitate degradation of ubiquitinated protein aggregates by autophagy. J Biol Chem 282 (33):24131–24145
6. Stolz A, Ernst A, Dikic I (2014) Cargo recognition and trafficking in selective autophagy. Nat Cell Biol 16(6):495–501
7. Jain A, Rusten TE, Katheder N, Elvenes J, Bruun JA, Sjottem E, Lamark T, Johansen T (2015) p62/Sequestosome-1, autophagy-related gene 8, and autophagy in Drosophila are regulated by nuclear factor erythroid 2-related factor 2 (NRF2), independent of transcription factor TFEB. J Biol Chem 290 (24):14945–14962
8. Nezis IP, Simonsen A, Sagona AP, Finley K, Gaumer S, Contamine D, Rusten TE, Stenmark H, Brech A (2008) Ref(2)P, the Drosophila melanogaster homologue of mammalian p62, is required for the formation of protein aggregates in adult brain. J Cell Biol 180 (6):1065–1071
9. Martinez-Lopez N, Athonvarangkul D, Singh R (2015) Autophagy and aging. Adv Exp Med Biol 847:73–87
10. Lystad AH, Simonsen A (2015) Assays to monitor aggrephagy. Methods 75:112–119

Methods in Molecular Biology (2018) 1854: 159–174
DOI 10.1007/7651_2018_122

Published online: 15 February 2018

Methods for Monitoring Autophagy in Silkworm Organs

Aurora Montali, Morena Casartelli, Daniele Bruno, Annalisa Grimaldi, and Gianluca Tettamanti

Abstract

In holometabolous insects, various larval organs are remodeled by autophagy during metamorphosis. Although moths and butterflies are among the first animal models in which this self-eating process was described, only in recent years autophagy has been analyzed in detail in these insects. In particular, the silkworm *Bombyx mori*, which represents a well-studied model among Lepidoptera, provides a wide repertoire of cellular and molecular tools useful for studying the occurrence of autophagy and for evaluating its role in postembryonic development. Here, we describe some morphological, biochemical, and molecular methods to monitor autophagy in silkworm organs.

Keywords Acid phosphatase, Atg8, Autophagy, *Bombyx mori*, Gene expression, Metamorphosis, Midgut, Silk gland, Transmission electron microscopy

1 Introduction

Metamorphosis represents a critical phase in the development of holometabolous insects. During the transition from the larval to the adult phase, several organs are fully remodeled or disappear as a result of apoptosis and autophagy. While apoptosis brings about cell death, the role of autophagy in this biological setting is quite controversial. In fact, since the 1960s, when Locke and Collins provided evidence on the occurrence of autophagy in *Calpodes ethlius* during metamorphosis [1, 2], numerous studies have analyzed this self-eating process in Lepidoptera. Accordingly, it has been proposed that autophagy promotes lipolysis in the larval fat body [3], while it has a prosurvival role in the larval midgut to provide trophic support to the epithelial cells [4, 5]. So far, however, these studies have not demonstrated a causative role of autophagy in cell death in larval organs of Lepidoptera. Among the main reasons for this delay in studying autophagy in these insects is probably the lack of a general agreement on the criteria to identify autophagic features and the availability of appropriate technical tools. This is a general problem in metazoans that, in the last 10 years, led to the publication of three versions of guidelines in which the standard criteria for monitoring autophagy in various eukaryotic models were defined [6–8]. While morphological analysis has been widely used in the past to confirm the occurrence of autophagy in Lepidoptera (see Ref. [9] for a review), the sequencing of *Bombyx mori* genome [10] and the development of new in vivo [3, 11, 12] and in vitro [13–15] tools in recent years have led to a significant advance in the knowledge of the autophagic processes in silkworm [16].

Here, we describe some techniques that can be used to routinely monitor autophagy in silkworm tissues.

1.1 Morphological Analysis

Although autophagy was initially described in Lepidoptera by using transmission electron microscopy (TEM) [9], this approach is often underutilized as it is time consuming and requires specific skills to analyze the results in order to avoid misinterpretations in identifying the autophagic compartments. However, not only different autophagic structures can be observed and identified by using this technique (Fig. 1a) but quantitative studies can also be performed by means of morphometric analyses.

1.2 Acid Phosphatase

Although the detection of lysosomes per se does not give sufficient evidence for assessing the activation of the autophagic process, as the degradation of the cargo sequestered inside the autophagosome is accomplished by fusion with lysosomes, the detection of lysosomal hydrolases can be used as a late marker of autophagy (Fig. 1b). In particular, it is possible to detect autolysosomes by using an ad hoc histochemical staining able to reveal acid phosphatase by TEM [17] (Fig. 1c, c′). In addition, the accumulation of lysosomes in tissues undergoing autophagy and a parallel increase in acid phosphatase levels have been observed in several silkworm organs [3, 17, 18].

1.3 Atg8 Lipidation and Expression of p4ebp1

The amount of lipidated Atg8/LC3 (Atg8-II) on western blot represents an undisputed indication of the accumulation of autophagosomes. In Lepidoptera, several custom or commercial antibodies have been used for this purpose [17, 19–22] but, at least in some cases, they proved to be inadequate. Starting from the evidence collected by Hu and colleagues [23] on the high similarity between silkworm Atg8 (BmAtg8) and Gabarap, we recently demonstrated that a commercial anti-Gabarap antibody is able to properly detect both Atg8 forms (i.e., BmAtg8 and BmAtg8-PE) on western blot [5] (Fig. 1d). Moreover, this antibody can detect BmAtg8-positive puncta in tissue samples by immunostaining [5].

Another protein marker to monitor the occurrence of autophagy is 4ebp1, a target of Torc1, which is phosphorylated upon activation of the autophagic pathway. Levels of phosphorylated 4ebp1 (p4ebp1) can be easily detected by western blotting [5].

1.4 Transcriptional Levels of ATG Genes

The induction of autophagy in the midgut [5, 17], silk gland [18, 24], and fat body [3] is accompanied by an increase in

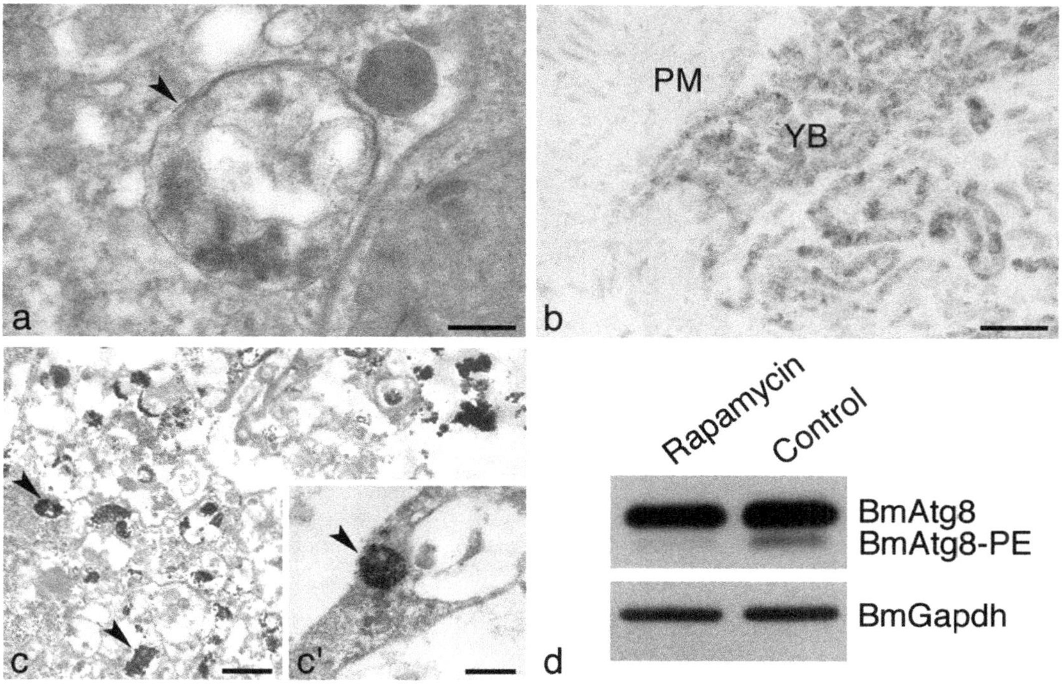

Fig. 1 (**a**) Transmission electron microscopy (TEM) image of an autophagosome characterized by a double-limiting membrane (arrowhead). (**b**, **c**) Acid phosphatase activity. The histochemical staining (**b**, red color) shows the localization of acid phosphatase activity in the yellow body (YB) cells that are degenerating within the newly forming pupal midgut (PM) epithelium. By TEM (**c**, black color), it is possible to observe the localization of acid phosphatase activity within autophagic compartments (arrowheads). (**d**) Western blot analysis of BmAtg8 in rapamycin-treated and control larvae. (**c′**) shows the detail at higher magnification of (**c**). *Bars*: 200 nm (**a**), 50 μm (**b**), 1 μm (**c**), and 500 nm (**c′**)

mRNA levels of some autophagic genes. This trend is common among insects [19, 25–27]. The transcriptional regulation of autophagy-related (*ATG*) genes in larval tissues during metamorphosis can be monitored by means of quantitative reverse transcription PCR (qRT-PCR), in which an appropriate housekeeping gene that is stable throughout development is used [5].

2 Materials

Prepare all the solutions with ultrapure water and analytical grade reagents. Use saline solution for Lepidoptera (SSL) to wash the organs isolated from the larva (sucrose 210 mM, KCl 45 mM, Tris–HCl 10 mM, pH 7.0). To prepare 100 mL: dissolve 0.121 g of Tris base (MW 121.14 g/mol) in 70 mL of water. Adjust pH to 7 with HCl. Add 0.335 g of KCl (MW 74.56 g/mol) and 7.188 g of sucrose (MW 342.30 g/mol), and bring to 100 mL with water.

2.1 Transmission Electron Microscopy Analysis

1. 0.2 M sodium cacodylate buffer, pH 7.4: dissolve 4.28 g of sodium cacodylate trihydrate (MW 214.03 g/mol) in 80 mL of water in a glass beaker. Adjust pH with HCl and bring to 100 mL with water. Store at 4 °C
2. 4% glutaraldehyde in 0.1 M sodium cacodylate buffer, pH 7.4: dilute 50% glutaraldehyde (commercial stock solution) to 8% glutaraldehyde with water. Dilute (1:1) the solution with 0.2 M sodium cacodylate buffer, pH 7.4. Store at 4 °C (*see* **Note 1**)
3. 1% osmium tetroxide in 0.1 M sodium cacodylate buffer, pH 7.4: dilute (1:1) 4% osmium tetroxide (commercial stock solution) with water and then dilute (1:1) the resulting solution with 0.2 M sodium cacodylate buffer, pH 7.4. Store at 4 °C in the dark
4. Ethanol series: 70% ethanol in water; saturated uranyl acetate in 90% ethanol (prepare 90% ethanol in water, then dissolve progressively the powder in 90% ethanol until saturation); 100% ethanol
5. Embedding resin (Epon-Araldite 812 mixture)
6. Propylene oxide/Epon-Araldite 812 mixture: mix (1:1) Epon-Araldite 812 and propylene oxide, and use immediately
7. Silicone embedding molds (Polysciences, Warrington, USA)
8. 200–300 mesh copper grids (Electron Microscopy Sciences, Hatfield, USA)
9. Lead citrate: weigh 1.33 g of lead nitrate (MW 331.2 g/mol) and 1.87 g of sodium citrate (MW 294.10 g/mol). Add 50 mL of water to a glass beaker and stir. When the powders are completely dissolved, add 8 mL NaOH 1 M. Store at 4 °C in the dark (*see* **Note 1**)
10. Saturated solution of uranyl acetate in water: dissolve progressively the powder in water until saturation. Store at 4 °C in the dark

2.2 Acid Phosphatase: Histochemical Staining

1. 10× PBS (1370 mM NaCl, 27 mM KCl, 100 mM $Na_2HPO_4 \times 12H_2O$, and 19.8 mM KH_2PO_4). To prepare 1 L: dissolve 80 g of NaCl (MW 58.44 g/mol), 2 g of KCl (MW 74.56 g/mol), 35.8 g of $Na_2HPO_4 \times 12H_2O$ (MW 358.14 g/mol), and 2.7 g of KH_2PO_4 (MW 136.09 g/mol) in 900 mL of water. Adjust pH to 7.4 and bring to 1 L with water. Autoclave the solution and dilute with water to obtain 1× PBS (working solution)
2. Freezing medium: prepare 20% sucrose (w/v) in 1× PBS. Mix Tissue-Tek® O.C.T. Compound (Sakura Finetek USA, Torrance, USA) with 20% sucrose (8:2) by using a vortex (*see* **Note 2**)

3. Tissue-Tek® Cryomold® molds, 10 × 10 × 5 mm (Sakura Finetek USA)
4. Polylysine- or gelatin-coated slides
5. Sodium acetate–acetic acid buffer solutions (50 mL): prepare 0.1 M CH_3COOH and 0.1 M CH_3COONa solutions. Mix 10.5 mL of 0.1 M CH_3COOH and 39.5 mL of 0.1 M CH_3COONa. Adjust pH to 5.2
6. Incubation mix: dissolve 10 mg of naphthol phosphate AS-BI in 2 mL of *N*-*N*-dimethylformamide. Mix 200 μL of this solution and 9.8 mL of sodium acetate–acetic acid buffer solution, pH 5.2. Add 6 mg of fast red violet LB and 6.3 μL of 10% $MnCl_2$ (w/v)
7. PBS/glycerol mounting medium: mix vigorously 1× PBS and glycerol (2:1)

2.3 Acid Phosphatase: Transmission Electron Microscopy

1. 0.2 M sodium cacodylate buffer, pH 7.4: see Subheading 2.1, point 1
2. 8% paraformaldehyde (PFA) (w/v) in distilled water: dissolve the powder on a hotplate stirrer (60 °C). When the powder is completely dissolved, cool the solution, filtrate with a paper filter, divide into aliquots, and stock at −20 °C until use
3. Karnovsky fixative (2% PFA and 2.5% glutaraldehyde (w/v) in 0.1 M sodium cacodylate buffer, pH 7.4). To prepare 50 mL: 12.5 mL 8% PFA (prepared as described in step 2), 2.5 mL 50% glutaraldehyde, 25 mL 0.2 M sodium cacodylate buffer, and 10 mL water. pH range 7.2 to 7.4 (*see* **Note 1**)
4. 1% osmium tetroxide in 0.1 M sodium cacodylate buffer, pH 7.4: see Subheading 2.1, point 3
5. 0.2 M Tris/maleate buffer, pH 5.2 (1 L): mix 24.2 g of tris (hydroxymethyl)aminomethane (MW 121.14 g/mol) and 23.2 g of maleic acid (MW 116.07 g/mol) in 1 L of water. Adjust pH at 5.2. Dilute to 40 mM Tris/maleate buffer (working solution)
6. 0.1 M sodium β-glycerophosphate
7. 0.02 M lead nitrate
8. Incubation medium (40 mM Tris/maleate buffer, 8 mM sodium β-glycerophosphate, and 2.4 mM lead nitrate) (50 mL): mix 10 mL of 0.2 M Tris/maleate buffer, 4 mL 0.1 M sodium β-glycerophosphate, 6 mL 0.02 M lead nitrate, and 30 mL of water

2.4 Acid Phosphatase Activity Assay

1. 1 M Tris base (MW 121.14 g/mol)
2. Homogenization buffer (100 mM mannitol, 10 mM HEPES-Tris, pH 7.2). To prepare 100 mL: dissolve 1.82 g of mannitol

and 0.24 g of HEPES in 100 mL of water. Adjust pH using 1 M Tris base. Add protease 100× inhibitor cocktail (1× final concentration) (Thermo Fisher Scientific, Waltham, USA) and 100× ethylenediaminetetraacetic acid (EDTA, Thermo Fisher Scientific) (1× final concentration)

3. Homogenizer (T10 basic ULTRA-TURRAX, IKA, Staufen, Germany)
4. Bradford reagent, 5× (Serva Electrophoresis GmbH, Heidelberg, Germany)
5. 45.9 mM citrate buffer (1 L): dissolve 13.5 g of sodium citrate tribasic dihydrate in 800 mL of water. Adjust pH to 4.9 with HCl. Bring to 1 L with water
6. 7.6 mM 4-nitrophenyl phosphate disodium salt hexahydrate in citrate buffer: mix 36 mg of 4-nitrophenyl phosphate disodium salt hexahydrate (MW 371.14 g/mol) in 12.7 mL of 45.9 mM citrate buffer
7. 0.1 M NaOH
8. White-walled 96-well plates (Bio-Rad, Hercules, USA)

2.5 Western Blot Analysis

1. 2.5 M sodium chloride
2. 1 M Tris–HCl, pH 8
3. RIPA buffer (150 mM sodium chloride, 2% NP-40, 0.5% sodium deoxycholate, 0.1% sodium dodecyl sulphate (SDS), 50 mM Tris, pH 8.0) (10 mL). First prepare 0.5% (w/v) sodium deoxycholate: mix 0.05 g of sodium deoxycholate (MW 432.59 g/mol) in 500 μL of water. Add 600 μL of 2.5 M sodium chloride, 200 μL of 100% NP-40, 50 μL 20% SDS, 500 μL of 1 M Tris base, pH 8, and 8.15 mL of water. Add 100× protease inhibitor cocktail (1× final concentration) (Thermo Fisher Scientific), phosphatase inhibitors (1 mM sodium orthovanadate, 5 mM sodium fluoride), (1× final concentration), and 100× EDTA (1× final concentration) (Thermo Fisher Scientific)
4. Homogenizer (T10 basic ULTRA-TURRAX, IKA)
5. Bradford reagent, 5× (Serva Electrophoresis GmbH)
6. Electrophoresis apparatus (Mini-PROTEAN® 3 Cell, Bio-Rad)
7. 1.5 M Tris–HCl, pH 8.8
8. 0.5 M Tris–HCl, pH 8.8
9. 12% running gel: 8.75 mL of water, 6 mL of 40% acrylamide/bis solution, 5 mL of 1.5 M Tris–HCl, pH 8.8, 100 μL of 20% SDS, 20 μL of *N*, *N*, *N*, *N′*-tetramethyl-ethylenediamine (TEMED), and 100 μL of 10% ammonium persulfate (APS)

10. 3.5% stacking gel: 5.50 mL of water, 700 μL of 40% acrylamide/bis solution, 2 mL of 0.5 M Tris–HCl, pH 8.8, 40 μL of 20% SDS, 9.6 μL of TEMED, and 60 μL of 10% APS
11. 2× Laemmli sample buffer (4% SDS, 10% 2-mercaptoethanol, 20% glycerol, 0.004% bromophenol blue, 0.125 M Tris–HCl, pH 6.8). To prepare 1 mL: mix 200 μL of 20% SDS, 100 μL of 2-mercaptoethanol, 230 μL of 87% glycerol, 5 μL of 2% (w/v) bromophenol blue, 250 μL of 0.5 M Tris–HCl, pH 6.8, and 215 μL of water
12. Prestained protein SHARPMASS™ VII, Protein MW marker (Euroclone, Pero, Italy)
13. 10× running buffer (1 L): dissolve 30.2 g of Tris base (MW 121.14 g/mol) and 144 g of glycine (MW 75.07 g/mol) in water
14. 1× running buffer (25 mM Tris base, 190 mM glycine, and 0.1% SDS) (1 L): mix 100 mL 10× running buffer with 895 mL of water, and add 5 mL of 20% SDS
15. Transfer container and transfer sandwich (Mini Trans-Blot® Electrophoretic Transfer Cell, Bio-Rad)
16. 10× blotting buffer (1 L): 24.6 g Tris base (MW 121.14 g/mol); 115 g glycine (MW 75.07 g/mol). Dissolve in water
17. 1× transfer buffer (1 L): 100 mL 10× blotting buffer, 200 mL methanol, and 700 mL water (*see* **Note 3**)
18. 0.45 μm PVDF (Merck Millipore, Burlington, Massachusetts, USA) and 0.45 μm nitrocellulose (Thermo Fisher Scientific, Waltham, Massachusetts, USA) membranes
19. Ponceau solution (0.1% (w/v) Ponceau S in 5% (v/v) trichloroacetic acid). Prepare 100 mL of 5% trichloroacetic acid in water and dissolve 0.1 g of Ponceau S (MW 760.6 g/mol)
20. TBS (1 L): 50 mL of 1 M Tris–HCl, pH 7.5, 60 mL 2.5 M NaCl, and 890 mL water
21. Blocking solution: 5% (w/v) nonfat milk in TBS
22. TBST: 0.1% (v/v) Tween20 in TBS
23. Antibodies: anti-Atg8 antibody (Abcam, Cambridge, UK, 1:3000 dilution), anti-p4ebp1 (Cell Signaling, Massachusetts, USA, 1:1000 dilution), anti-GAPDH (Proteintech, Rosemont, USA, 1:2500 dilution), and anti-rabbit IgG HRP conjugate (The Jackson Laboratory, Bar Harbor, USA, 1:7500 dilution). The working solution must be prepared by diluting the antibody with 5% milk in TBS (*see* **Note 4**)

2.6 Quantitative Reverse Transcription PCR

1. Homogenizer (T10 basic ULTRA-TURRAX, IKA)
2. Trizol reagent (Life Technologies, Carlsbad, USA)
3. Isopropanol

4. Chloroform
5. RNase-free water
6. 70% Ethanol in RNA-free water
7. TURBO DNA-free Kit (Life Technologies)
8. 1× Tris-acetate-EDTA (TAE) (Euroclone)
9. 1% agarose gel (100 mL): dissolve 1 g of agarose in 100 mL of 1× TAE. Put in a microwave for 1–3 min until the agarose is completely dissolved. Allow agarose solution to cool down and add 5 μL of Eurosafe-Fluorescent Nucleic Acid Stain 20,000× (Euroclone). Pour the agarose into a gel casting tray with the well comb
10. 10 mM dNTP mix (Euroclone)
11. Oligo $(dT)_{12-18}$ (500 μg/mL)
12. RNaseOUT™ Recombinant Ribonuclease Inhibitor (Thermo Fisher Scientific)
13. M-MLV Reverse Transcriptase (Life Technologies)
14. iTaq Universal SYBR Green Supermix (Bio-Rad)
15. 96-well CFX Connect Real-Time PCR Detection System (Bio-Rad)

3 Methods

The methods described below have been used to monitor autophagy in the larval midgut [5, 17] and posterior silk gland [18] of *B. mori*. To isolate the organs, anesthetize the larvae with CO_2 for a few seconds. Then, dissect the larva dorsally by cutting the integument and remove the target organs (Fig. 2). Before further processing, wash thoroughly the larval sample with SSL to remove

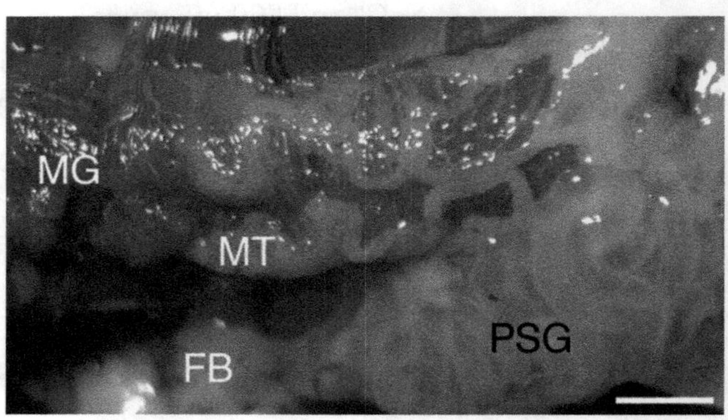

Fig. 2 Stereomicroscopy image showing the internal anatomy of the silkworm. The anterior part of the larva is on the left. *FB* fat body, *PSG* posterior silk gland, *MG* midgut, *MT* Malpighian tubules. *Bar*: 2 mm

contaminants (hemocytes and tissue debris). If the midgut is the selected organ and the method requires homogenization of the tissue (acid phosphatase activity assay, western blot analysis, and qRT-PCR) (*see* Subheadings 3.4, 3.5, and 3.6), Malpighian tubules and peritrophic matrix with its content must be removed; then, wash the sample with SSL.

3.1 Transmission Electron Microscopy Analysis

Process the samples in 1.5-mL plastic tubes. Use at least 1 mL of the required solution, unless otherwise specified.

1. Fix samples in 4% glutaraldehyde at 4 °C overnight (*see* **Note 5**).
2. Wash in 0.1 M cacodylate buffer (5 washes, 5 min each).
3. Postfix specimens with 0.5 mL of 1% osmium tetroxide in 0.1 M cacodylate buffer, pH 7.4 for 1 h (*see* **Note 6**).
4. Wash in 0.1 M cacodylate buffer (5 washes, 5 min each).
5. Dehydrate samples in a graded series of cold ethanol: 70% ethanol for 30 min (*see* **Note 7**), saturated uranyl acetate in 90% ethanol for 30 min (*see* **Note 6**), and twice with 100% ethanol for 15 min each.
6. Incubate tissues in propylene oxide/Epon-Araldite 812 mixture for 1 h (*see* **Note 1**).
7. Discard the mixture and allow the propylene oxide residues to evaporate from the tube, leaving it open under the fume hood for some minutes (*see* **Note 8**).
8. Add Epon-Araldite 812 mixture to the samples and let them stand overnight at room temperature.
9. Transfer the samples to embedding molds; cover them with fresh resin and allow them to polymerize overnight at 70 °C.
10. Cut thin sections (70-nm-thick) with an ultramicrotome and collect them on copper grids.
11. Stain sections: 4 min lead citrate, 10 washes with water, 8 min uranyl acetate, 10 washes with water, 4 min lead citrate, and 10 washes with water (*see* **Note 9**).

3.2 Acid Phosphatase: Histochemical Staining

1. Dissect the larva and isolate the organs (see the introduction paragraph of Subheading 3).
2. Put the sample in cryomolds and cover them completely with freezing medium; freeze with liquid nitrogen and store at −80 °C until use.
3. Cut cryosections (8-μm-thick) with a cryostat and use the slides immediately or store at −20 °C (*see* **Note 10**).
4. Wash cryosections with 1× PBS for 5 min.
5. Add sodium acetate–acetic acid buffer solutions, pH 5.2, for 5 min.

6. Incubate with the incubation mix for 90 min at 37 °C (*see* **Note 11**).
7. Wash with 1× PBS (3 washes, 5 min each).
8. Mount coverslip with PBS/glycerol mounting medium.

3.3 Acid Phosphatase: Transmission Electron Microscopy

1. Isolate organs from the larva (see the introduction paragraph of Subheading 3).
2. Fix samples with Karnovsky fixative for 2 h at 4 °C (*see* **Note 5**).
3. Wash in 0.1 M cacodylate buffer (2 washes, 15 min each).
4. Wash with 40 mM Tris/maleate buffer for 10 min.
5. Incubate with incubation medium for 2 h at 37 °C (*see* **Note 11**).
6. Wash with 40 mM Tris/maleate buffer (2 washes, 15 min each).
7. Postfix specimens with 0.5 mL of 1% osmium tetroxide in 0.1 M cacodylate buffer, pH 7.4 for 1 h (*see* **Note 6**).
8. Wash in 0.1 M cacodylate buffer (5 washes, 5 min each).
9. Complete the processing procedure as indicated in Subheading 3.1 (points 5–11).

3.4 Acid Phosphatase Activity Assay

1. Dissect the larva and isolate organs (see the introduction paragraph of Subheading 3).
2. Homogenize samples with homogenization buffer (10 μL buffer/μg tissue) maintaining the tube on ice.
3. Shake the tube for 1 h on ice.
4. Centrifuge the tube at 15,000 × *g* for 10 min at 4 °C.
5. Transfer the supernatant into a new tube.
6. Determine protein concentration by using the Bradford assay (*see* Ref. [28]).
7. Prepare the reaction mix: put 250 μL of 7.6 mM 4-nitrophenyl phosphate in citrate buffer (45.9 mM, pH 4.9) into a 1.5-mL plastic tube and add different volumes of homogenate samples (i.e., 2, 4, and 6 μL) diluted to 50 μL in homogenization buffer into different wells. Prepare for each sample a blank reaction mix without the substrate: put 250 μL of citrate buffer (45.9 mM, pH 4.9) into different plastic tubes and add the same volumes of homogenate used for samples, i.e., 2, 4, and 6 μL, diluted to 50 μL in homogenization buffer) (*see* Ref. [29]).
8. Incubate at 25 °C for 30 min.
9. Stop the reaction with 1 mL of 0.1 M sodium hydroxide.
10. Put 300 μL of the mix into a white-walled 96-well plate.

11. Determine color development at 405 nm using an Infinite F200 96-well plate reader. Subtract blank values from the corresponding sample values.
12. To calculate the enzyme activity from the absorbance value, *see* Ref. [29].

3.5 Western Blot Analysis

1. Dissect the animal and remove the target organ (see the introduction paragraph of Subheading 3); freeze samples in liquid nitrogen and store at −80 °C until use.
2. Transfer the sample to a 2-mL plastic tube.
3. Add RIPA buffer and homogenize the sample maintaining the tube on ice (*see* **Note 12**).
4. Put the tube in ice and shake for 1 h.
5. Centrifuge the tube at 15,000 × *g* for 15 min at 4 °C.
6. Transfer the supernatant to a new tube (*see* **Note 13**).
7. Quantify the protein content according to the Bradford method (*see* Ref. [28]).
8. Prepare vertical electrophoresis apparatus.
9. Pour 12% running gel and add 1 mL isopropanol.
10. When the running gel is completely solidified, drain the isopropanol and add 3.5% stacking gel. Insert a 10-well gel comb immediately, avoiding the formation of air bubbles.
11. When the stacking gel is completely solidified, remove the gel comb.
12. Insert gel in the electrophoresis apparatus and cover with 1× running buffer.
13. Load 40 μg of protein into each well and 5 μL of protein MW marker.
14. Run the gel at 50 V for 40 min (until the dye front passes the stacking gel), then at 100 V for 1–2 h until the dye front reaches the bottom of the gel.
15. When the SDS-PAGE is completed, cut off the stacking gel.
16. Lay the membrane on the running gel and assemble the transfer sandwich (*see* Ref. [30]) (*see* **Note 14**).
17. Place the sandwich in the transfer container and fill it with cold transfer buffer. Run for 2 h (constant current) (*see* **Note 15**).
18. To visualize the correct transfer of the proteins, stain the membrane with Ponceau solution (*see* **Note 16**) and then wash it with water. Dry the membrane (*see* **Note 17**).
19. Block the membrane by soaking it in blocking solution for 1 h with shaking (*see* **Note 18**).
20. Incubate with primary antibody for 2 h with shaking.

21. Wash 3 times with TBST (10 min each).
22. Incubate with anti-rabbit IgG HRP conjugate for 1 h with shaking.
23. Wash 3 times with TBST (10 min each).
24. Wash once with TBS for 10 min.
25. Incubate for 5 min with SuperSignal West Pico Substrate (Thermo Fisher Scientific) and expose blot to film for 5 s–5 min.

3.6 Quantitative Reverse Transcription PCR

1. Dissect larvae and isolate organs (see the introduction paragraph of Subheading 3); freeze samples in liquid nitrogen and store at −80 °C until use.
2. Transfer specimens to 2-mL plastic tubes.
3. Add 1 mL of Trizol reagent and thoroughly homogenize the sample while keeping the tube on ice.
4. Incubate the homogenized sample at room temperature for 5 min.
5. Add 200 μL of cold chloroform keeping the sample on ice and then shake the tube vigorously for 15 s.
6. Incubate at room temperature for 3 min.
7. Centrifuge the tube at 12,000 × *g* for 15 min at 4 °C.
8. Remove 500 μL of aqueous phase and place in a new tube.
9. Add 50 μL of cold 100% isopropanol and incubate at room temperature for 5 min.
10. Centrifuge at 12,000 × *g* for 10 min at 4 °C.
11. Remove 500 μL of aqueous phase and place into a new tube.
12. Add 310 μL of 100% isopropanol, invert the tube properly, and incubate at room temperature for 10 min.
13. Centrifuge at 12,000 × *g* for 10 min at 4 °C.
14. Completely remove the supernatant.
15. Wash the pellet with cold 75% ethanol and vortex briefly.
16. Centrifuge at 7500 × *g* for 5 min at 4 °C.
17. Remove ethanol and allow the pellet to dry by leaving the tube open (*see* **Note 19**).
18. Resuspend the RNA pellet with 30–150 μL of RNase-free water; incubate in a heat block at 55 °C for 10 min to allow complete resuspension of the pellet.
19. Quantify total RNA with a spectrophotometer, divide into aliquots, and store at −80 °C until use.

20. For treatment with TURBO DNA-free Kit, dilute 10 μg of total RNA in 26 μL of RNase-free water.
21. Treat RNA with TURBO DNA-free Kit (Life Technologies).
22. Quantify RNA with a spectrophotometer and analyze the integrity of RNA by running a small aliquot (1 μg) on 1% agarose gel electrophoresis.
23. For reverse transcription, dilute 2 μg of RNA in 10 μL of RNase-free water. Add 2 μL of mix 1 (see Table 1) and incubate for 5 min at 65 °C and quickly cool down on ice. After a quick centrifugation, add 7 μL of mix 2 (see Table 1) and incubate for 2 min at 37 °C. Add 1 μL of M-MLV reverse transcriptase (Life Technologies), incubate at 37 °C for 50 min, and stop the reaction by incubating the mix at 70 °C for 15 min.
24. Dilute (1:8) cDNA with autoclaved ultrapure water.
25. Perform qRT-PCR by using a 96-well CFX Connect Real-Time PCR Detection System. The list of primers and the PCR reaction mix are reported in Tables 2 and 3, respectively. Cycling conditions for real-time PCR: 95 °C for 30 s, 95 °C for 10 s (39 cycles), and 61 °C for 30 s.

Table 1
Reverse transcription reaction mix

	Mix component
Mix 1	1 μL Oligo $(dT)_{12-18}$ (500 μg/mL) 1 μL 10 mM dNTP mix
Mix 2	4 μL 5× First-strand buffer 2 μL M DTT 0.1 mM 1 μL RNaseOUT™ Recombinant Ribonuclease Inhibitor

Table 2
Primer sequences

Gene	GenBank number	Primer sequences
BmATG1	NM_001309546.1	F: CCCCGCCTATGTCTATGTTG R: ATCTGATGGGTGGGAGTACG
BmATG8	NM_001046779.1	F: CCAGATCGCGTTCCTGTAAT R: GAGACCCCATTGTTGCAGAT
BmRP49	NM_001098282.1	F: AGGCATCAATCGGATCGCTATG R: TTGTGAACTAGGACCTTACGGAATC

Table 3
Real-time PCR reaction mix

Reagent	Final concentration
iTaq Universal SYBR Green Supermix	1×
Forward primer	0.2 μM
Reverse primer	0.2 μM
Ultrapure water	Up to 20 μL

4 Notes

1. This solution must be freshly prepared each time.
2. Prepare the solution at least one day before use to allow air bubbles to disappear. Store at 4 °C to avoid mold formation.
3. This solution must be freshly prepared each time and cooled down at −20 °C for at least 2 h before use.
4. To detect both BmAtg8 forms, we suggest using an anti-Gabarap antibody (Abcam, Cambridge, UK, # AB109364).
5. To obtain an appropriate fixation of the tissue, process samples of reduced size, use at least 1 mL of fixative/sample, and fix the sample under gentle agitation.
6. The samples must be kept in the dark during incubation.
7. At this stage, the processing procedure can be interrupted, if necessary. The samples can be stored in 70% ethanol at 4 °C until further processing.
8. 5–10 min should be sufficient. Please consider that residues of propylene oxide inside the samples soften the embedding resin and compromise sectioning.
9. All the solutions must be filtered with a 0.2-μm syringe filter before use.
10. Air-dry cryosections accurately before use. Moreover, collect sections on polylysine- or gelatin-coated slides. These two precautions will promote the adhesion of cryosections to the slide.
11. The incubation procedure must be performed in a humid chamber to avoid evaporation of the incubation mix.
12. To improve protein extraction, use a different amount of tissue versus RIPA buffer (mg of tissue/μL RIPA buffer), i.e., 1:5 for silk gland and 1:10 for midgut samples.
13. If the supernatant is not clear, centrifuge the tube again. The protein sample can be stored at −80 °C until use for SDS-PAGE.

14. Use a 0.45-μm PVDF membrane for Atg8 detection and a 0.45-μm nitrocellulose membrane for p4ebp1; the PVDF membrane must be activated by immersion in methanol for 20 s, ultrapure water for 2 min, and transfer buffer for 5 min.
15. Running amperage is different according to the number of gels inside the transfer container: 350 mA for one gel; 450 mA for two gels.
16. PVDF membranes can be stained by using a Ponceau:methanol solution (1:1) for 5 min.
17. The membrane can be stored at 4 °C until use for immunoblotting.
18. PVDF membrane must be first activated by immersion in methanol for 20 s and drying it. Then, blocking solution can be added.
19. Three minutes are usually sufficient to dry the RNA pellet without impairing the subsequent resuspension in water. This step must not be performed under a fume hood.

Acknowledgements

This work was partially supported by FAR 2017 (University of Insubria) to GT. Aurora Montali is a Ph.D. student of the "Life Sciences and Biotechnology" course at Università degli Studi dell'Insubria. Daniele Bruno is a Ph.D. student of the "Biotechnologies, Biosciences and Surgical Technologies" course at Università degli Studi dell'Insubria.

References

1. Locke M, Collins JV (1965) The structure and formation of protein granules in the fat body of an insect. J Cell Biol 26:857–884
2. Locke M, Collins JV (1968) Protein uptake into multivesicular bodies and storage granules in the fat body of an insect. J Cell Biol 36:453–483
3. Tian L, Ma L, Guo E, Deng X et al (2013) 20-hydroxyecdysone upregulates Atg genes to induce autophagy in the *Bombyx* fat body. Autophagy 9:1172–1187
4. Franzetti E, Romanelli D, Caccia S, Cappellozza S, Congiu T et al (2015) The midgut of the silkmoth *Bombyx mori* is able to recycle molecules derived from degeneration of the larval midgut epithelium. Cell Tissue Res 361:509–528
5. Romanelli D, Casartelli M, Cappellozza S, de Eguileor M, Tettamanti G (2016) Roles and regulation of autophagy and apoptosis in the remodelling of the lepidopteran midgut epithelium during metamorphosis. Sci Rep 6:32939
6. Klionsky DJ, Abeliovich H, Agostinis P, Agrawal DK, Aliev G et al (2008) Guidelines for the use and interpretation of assays for monitoring autophagy in higher eukaryotes. Autophagy 4:151–175
7. Klionsky DJ, Abdalla FC, Abeliovich H, Abraham RT, Acevedo-Arozena A et al (2012) Guidelines for the use and interpretation of assays for monitoring autophagy. Autophagy 8:445–544
8. Klionsky DJ, Abdelmohsen K, Abe A, Abedin MJ, Abeliovich H et al (2016) Guidelines for

the use and interpretation of assays for monitoring autophagy (3rd edition). Autophagy 12:1–222

9. Romanelli D, Casati B, Franzetti E, Tettamanti G (2014) A molecular view of autophagy in lepidoptera. Biomed Res Int 2014:902315
10. Mita K, Kasahara M, Sasaki S, Nagayasu Y, Yamada T (2004) The genome sequence of silkworm, *Bombyx mori*. DNA Res 11:27–35
11. Liu X, Dai F, Guo E, Li K, Ma L et al (2015) 20-Hydroxyecdysone (20E) primary response gene E93 modulates 20E signaling to promote *Bombyx* larval-pupal metamorphosis. J Biol Chem 290:27370–27383
12. Hu W, Liu C, Cheng T, Li W, Wang N, Xia Q (2016) Histomorphometric and transcriptomic features characterize silk glands' development during the molt to intermolt transition process in silkworm. Insect Biochem Mol Biol 76:95–108
13. Ji MM, Lee JM, Mon H, Xu J, Tatsuke T, Kusakabe T (2016) Proteasome inhibitor MG132 impairs autophagic flux through compromising formation of autophagosomes in *Bombyx* cells. Biochem Biophys Res Commun 479:690–696
14. Ji MM, Lee JM, Mon H, Iiyama K, Tatsuke T et al (2017) Lipidation of BmAtg8 is required for autophagic degradation of p62 bodies containing ubiquitinated proteins in the silkworm, *Bombyx mori*. Insect Biochem Mol Biol 89:86–96
15. Xie K, Tian L, Guo X, Li K, Li J et al (2016) BmATG5 and BmATG6 mediate apoptosis following autophagy induced by 20-hydroxyecdysone or starvation. Autophagy 12:381–396
16. Xia Q, Li S, Feng Q (2014) Advances in silkworm studies accelerated by the genome sequencing of *Bombyx mori*. Annu Rev Entomol 59:513–536
17. Franzetti E, Huang ZJ, Shi YX, Xie K, Deng XJ et al (2012) Autophagy precedes apoptosis during the remodeling of silkworm larval midgut. Apoptosis 17:305–324
18. Montali A, Romanelli D, Cappellozza S, Grimaldi A, de Eguileor M, Tettamanti G (2017) Timing of autophagy and apoptosis during posterior silk gland degeneration in *Bombyx mori*. Arthropod Struct Dev 46:518–528
19. Khoa DB, Takeda M (2012) Expression of autophagy 8 (Atg8) and its role in the midgut and other organs of the greater wax moth, *Galleria mellonella*, during metamorphic remodelling and under starvation. Insect Mol Biol 21:473–487
20. Gai Z, Zhang X, Islam M, Wang X, Li A et al (2013) Characterization of Atg8 in lepidopteran insect cells. Arch Insect Biochem Physiol 84:57–77
21. Li YB, Li XR, Yang T, Wang JX, Zhao XF (2016) The steroid hormone 20-hydroxyecdysone promotes switching from autophagy to apoptosis by increasing intracellular calcium levels. Insect Biochem Mol Biol 79:73–86
22. Shiba H, Yabu T, Sudayama M, Mano N, Arai N et al (2016) Sequential steps of macroautophagy and chaperone-mediated autophagy are involved in the irreversible process of posterior silk gland histolysis during metamorphosis of *Bombyx mori*. J Exp Biol 219:1146–1153
23. Hu C, Zhang X, Teng YB, Hu HX, Li WF (2010) Structure of autophagy-related protein Atg8 from the silkworm *Bombyx mori*. Acta Crystallogr Sect F Struct Biol Cryst Commun 66:787–790
24. Li Q, Deng X, Yang W, Huang Z, Tettamanti G et al (2010) Autophagy, apoptosis, and ecdysis-related gene expression in the silk gland of the silkworm (*Bombyx mori*) during metamorphosis. Can J Zool 88:1169–1178
25. Denton D, Shravage B, Simin R, Mills K, Berry DL et al (2009) Autophagy, not apoptosis, is essential for midgut cell death in *Drosophila*. Curr Biol 19:1741–1746
26. Santos DE, Azevedo DO, Campos LA, Zanuncio JC, Serrão JE (2015) *Melipona quadrifasciata* (Hymenoptera: Apidae) fat body persists through metamorphosis with a few apoptotic cells and an increased autophagy. Protoplasma 252:619–627
27. Tindwa H, Jo YH, Patnaik BB, Lee YS, Kang SS, Han YS (2015) Molecular cloning and characterization of autophagy-related gene TmATG8 in Listeria-invaded hemocytes of *Tenebrio molitor*. Dev Comp Immunol 51:88–98
28. Bradford MM (1976) A rapid and sensitive method for the quantitation of microgram quantities of protein utilizing the principle of protein-dye binding. Anal Biochem 72:248–254
29. Moss WD (1983) Methods of enzymatic analysis. In: Bergmeyer J, Grassi M (eds) Esterases, glycosidases, lyases, ligases, vol 4. Verlag-Chemie, Weinheim, pp 92–106
30. Towbin H, Staehelin T, Gordon J (1979) Electrophoretic transfer of proteins from polyacrylamide gels to nitrocellulose sheets: procedure and some applications. Proc Natl Acad Sci U S A 76:4350–4354

Methods in Molecular Biology (2018) 1854: 175–185
DOI 10.1007/7651_2018_124

Published online: 27 March 2018

Immunofluorescence Staining Protocols for Major Autophagy Proteins Including LC3, P62, and ULK1 in Mammalian Cells in Response to Normoxia and Hypoxia

Wen Li, Shupeng Li, Yifang Li, Xiaoying Lin, Yongquan Hu, Tian Meng, Baojin Wu, Rongrong He, and Du Feng

Abstract

Immunofluorescence is an invaluable technique widely used in cell biology. This technique allows visualization of the subcellular distribution of different target proteins or organelles, by specific recognition of the antibody to the endogenous protein itself or to its antigen via the epitope. This technique can be used on tissue sections, cultured cells, or individual cells. Meanwhile, immunofluorescence can also be used in combination with non-antibody fluorescent staining, such as DAPI or fluorescent fusion proteins, e.g., GFP or YFP, etc.

Autophagy is a catabolic pathway in which dysfunctional organelles and cellular components are degraded via lysosomes. During this process, cytoplasmic LC3 translocates to autophagosomal membranes. Therefore, cells undergoing autophagy can be identified by visualizing fluorescently labeled LC3 or other autophagy markers. Immunofluorescence is an important part of autophagy detection methods even if observation of the formation of autophagosome by transmission electron microscopy has become a gold standard for characterizing autophagy.

By observing the immunofluorescence staining of some key autophagy proteins, we can intuitively evaluate the levels of autophagy in samples. Herein, this protocol describes the predominant method used for the research of autophagy, which mainly focuses on the immunofluorescence staining of cellular LC3, P62, and ULK1 in response to normoxia and hypoxia, by presenting the detailed materials required and methodology.

Keywords Autophagy, Immunofluorescence, LC3, Mitophagy, P62, ULK1

1 Introduction

Autophagy is a conserved and tightly controlled intracellular degradation pathway to secure cell homeostasis by eliminating aggregated proteins and damaged organelles [1]. This central cellular mechanism plays significant roles in cellular remodeling during differentiation and development [2, 3], and is also responsible for the occurrence and development of many diseases, for example, cancer [4–6], neurodegenerative diseases [7, 8], inflammatory reaction [9, 10], virus infections [11, 12], and so on.

ULK1 (unc-51 like autophagy activating kinase 1), a mammalian homolog of Atg1, regulates the induction of autophagy in response to starvation or MTORC1 [mechanistic target of rapamycin (serine/threonine kinase) complex 1] inhibition. ULK1 is phosphorylated by MTORC1 and AMPK (5′ AMP-dependent kinase). Activation of ULK1 by AMPK-mediated phosphorylations is required for the induction of autophagy. However, MTORC1 inhibits autophagy by phosphorylating and negatively regulating ULK1 and its binding partner, ATG13 [13–16]. Meanwhile, functioning as a conserved serine–threonine kinase, ULK1 can also bind and phosphorylate ATG14 under nutrient deprivation or MTORC1 inhibition. This regulation is momentous for phagophore and autophagosome formation [17].

P62, also known as sequestosome1, is a multidomain adaptor protein [18]. By interacting selectively with different signaling proteins, P62 is involved in multiple cellular functions including the activation of NF-κB [19, 20], extrinsic apoptosis [21–23], tumorigenesis [24], and autophagy [25, 26]. Functioning as a key regulator of nutrient sensing in the mTORC1 pathway, it has been revealed that P62 colocalizes and interacts with Rags and is required for translocation of the mTORC1 to the lysosome, a crucial step for mTOR activation [27]. In addition, P62 binds LC3 through an LC3-interacting region (LIR).

Autophagy is tightly regulated by autophagy-related genes (ATG). Among these genes, LC3 (a mammalian homolog of Atg8) is required for autophagosome maturation and selective autophagy [28]. The C-terminal polypeptide of LC3 is cleaved by ATG4 protease to expose a glycine residue. Cleaved LC3 is then covalently conjugated to phosphatidylethanolamine (PE) by interacting with ATG7, ATG3 [29], and ATG12–ATG5–ATG16L [30, 31] complex (ubiquitin-like modification process), to generate LC3-PE (referred to as LC3-II) that binds to the autophagic membrane.

Phosphorylation of LC3 at threonine 50 (Thr50) by STK3 (serine/threonine kinase 3) and STK4 (serine/threonine kinase 4) is essential for autophagy, that is because loss of this phosphorylation blocks autophagy by impairing fusion of autophagosomes with lysosomes [32]. LC3 is also an autophagic substrate due to its degradation by autophagy. Consequently, LC3 has been widely used to monitor the number of autophagosomes as well as autophagic activity.

Endogenous ULK1 cellular localization is analyzed using ULK1 immunofluorescence in MEFs and HeLa cells. ULK1 puncta mainly distribute in cytoplasm. A greater number of ULK1 puncta in cells are induced in autophagic cells under different stress condition. Immunofluorescence staining of P62 reveals a clear, punctate pattern localized into cytoplasm. LC3, another autophagy biomarker except for P62, also diffuses evenly in cytoplasm in the absence of autophagy. However, once autophagy is initiated, aggregated LC3 puncta are formed. A plurality of bright spots is observed under a fluorescent microscope. A spot is equivalent to an autophagosome, and the activity of autophagy can be evaluated by counting the numbers of spots. However, the increased number of LC3 spots does not necessarily represent an increase in autophagy activity. It is also possible that autophagy is blocked by lysosomal degradation pathways, which needs to be further confirmed by western blot for detecting the expression level of P62 (an autophagy degrading substrate) and electron microscopy. If mitophagy occurs, LC3 will translocate from cytoplasm to enclose mitochondria. Hence, co-localization of fluorescently labeled LC3 and mitochondrial markers can be visualized.

2 Materials

2.1 Required Materials, Solutions, and Regents

Slides (CITOGLAS, 10127105P-G)

Antifade Mounting Medium (Beyotime, P0128)

Nail enamel, Coverslips (Sangon Biotech, F518113)

Moist environmental box

Triton X-100 (MBCHEM, T0694)

Phosphate buffered saline, PBS (Solarbio, P1010)

Albumin from bovine serum, BSA (Biosharp, BS043B)

4% Paraformaldehyde, PFA (BBI Life Sciences, E672002-0500), anti-ULKl (Sigma-Aldrich, A7481)

Anti-LC3B (Sigma, L7543)

Anti-LC3 (MBL, M152-3, PM036)

Anti-LC3 (NOVUS, NB100-2220)

Anti-SQSTM1/P62 (Abcam, ab56416)

Control IgG, 4′, 6-diamidino-2-phenylindole, DAPI (CST, 4083)

Alexa fluor 555-labeled donkey anti-mouse IgG antibodies (Invitrogen, A31570)

Alexa fluor 488-labeled donkey anti-rabbit IgG antibodies (Invitrogen, A21206)

PBS solution: 0.01 M, pH 7.2–7.4, store at 4 °C

Fixative solution: 4% PFA, store at 4 °C

Permeabilizing solution: 0.1% (V/V) triton X-100 in PBS solution, store at room temperature

Blocking solution buffer: 0.1 g BSA in 10 mL PBS. Temporarily prepared and used

Primary antibody diluent: 0.01% (V/V) triton X-100 in PBS solution, store at room temperature

Secondary antibody diluent: 1% BSA in PBS. Temporarily prepared and used

All solutions were prepared with ultrapure water (prepared by purifying deionized water to attain a sensitivity of 18 MΩ cm at 25 °C)

2.2 Coverslip Preparation

Coverslips (size, 22 mm × 22 mm; thickness, 0.13–0.17 mm) are immersed in the cleaning liquid (kalium bichromicum:concentrated sulfuric acid:distilled water = 120 g:200 mL:200 mL) overnight. Take out the coverslips and rinse coverslips with tap water until there is no cleaning liquid residue. Use ultrapure water to rinse the coverslip again for 10 times. Subsequently, coverslips are soaked in 75% ethanol solution for the following experiments. Glass bottom dish, a 35-mm dish with 14-mm bottom well (1 glass, 0.13–0.16 mm), is more commonly used for cells difficult to adhere or observing living cells.

3 Methods

3.1 Indirect Immunofluoresence

Carry out all procedures at room temperature unless otherwise specified. Immunofluorescence staining protocols of adherent cells are described as follows:

1. Clip a coverslip from 75% ethanol solution, and dry the coverslip by spirit lamp. Then, leave the coverslip in 6-well culture plate or 35 mm cell culture dish. This step can be omitted if a glass bottom dish is used.
2. Grow cultured cells on coverslip till 60–80% confluence.
3. Remove half of the medium, and add the same volume of 4% PFA in the petri dish to incubate for 5–10 min at 37 °C.
4. Discard the supernatant and add 1 mL 4% PFA in the petri dish again, followed by 15 min incubation at 37 °C.
5. Remove solution and rinse cells with PBS for 3 × 5 min.
6. Discard PBS and add 1 mL permeabilizing solution for 15–20 min at room temperature (*see* **Note 1**).
7. Remove solution and rinse cells with PBS for 3 × 5 min.
8. Add 1 mL blocking solution buffer, and leave for 30 min at room temperature (*see* **Note 2**).
9. Discard the solution, and incubate the primary antibody onto coverslip for 1 h at room temperature or overnight at 4 °C in moist environmental box (*see* **Note 3**). Primary antibody is diluted by primary antibody diluent (*see* **Notes 4** and **5**). The two primary antibodies are mixed and incubated together for double immunofluorescence staining. Nevertheless, in some cases, it needs addition sequentially.

10. Rewarm the wet box at room temperature for 1 h. Reclaim and preserve the primary antibody into 1.5 mL tube at indicated condition. Rinse cells with PBS for 3 × 5 min.
11. The fluorophore labeled secondary antibodies are diluted into secondary antibody diluent (*see* **Notes 6** and 7). Incubate the secondary antibody onto coverslip for 1 h at room temperature in moist environmental box. The two secondary antibodies are mixed and incubated together for double immunofluorescence staining. However, in some cases, it needs addition sequentially. Carry out the following procedures in dark place unless otherwise specified.
12. Discard the solution and rinse cells with PBS for 3 × 5 min.
13. If necessary, sometimes we also stain DAPI to help us get a better sense of cell population and separate the morphology and location of cytoplasm and nucleus region. DAPI is diluted into PBS in accordance with the instructions and added onto coverslip for 5 min at room temperature.
14. Remove solution and rinse cells with PBS for 3 × 5 min.
15. 20 μL of mounting medium is promptly added onto the slide, then put a cover slip on it, followed by mounting the coverslip with nail enamel.
16. Preserve the coverslip at 4 °C, in dark place.

3.2 Image Acquisition

In each experiment, cell images are captured with a TCS SPF5 II Leica confocal microscope (LAS-AF-Lite_2.2.0_4758; Leica Microsystems, Wetzlar, Germany). The relative average areas of autophagosomes can be calculated by using Image-Pro Plus 6.0 according to the tutorial.

4 Notes

1. Triton X-100 ($C_{14}H_{22}O(C_2H_4O)n$), a nonionic surfactant that has a hydrophilic polyethylene oxide chain and an aromatic hydrocarbon lipophilic or hydrophobic group, can dissolve the lipid cellular membrane and nuclear membrane by forming a hole to let the antibodies into cytoplasm or nucleus [33, 34]. Therefore, triton X-100 is also called drilling agent. Generally use the concentration of 0.1% (V/V).
2. For some antibody that has less specificity, the time for block can be appropriate to extend to 1 h.
3. In order to retrench the use of antibodies, generally speaking, 100 μL solution can be completely covered with a piece of coverslip. For glass bottom dish, 50 μL solution is already enough to cover the entire range of cells.

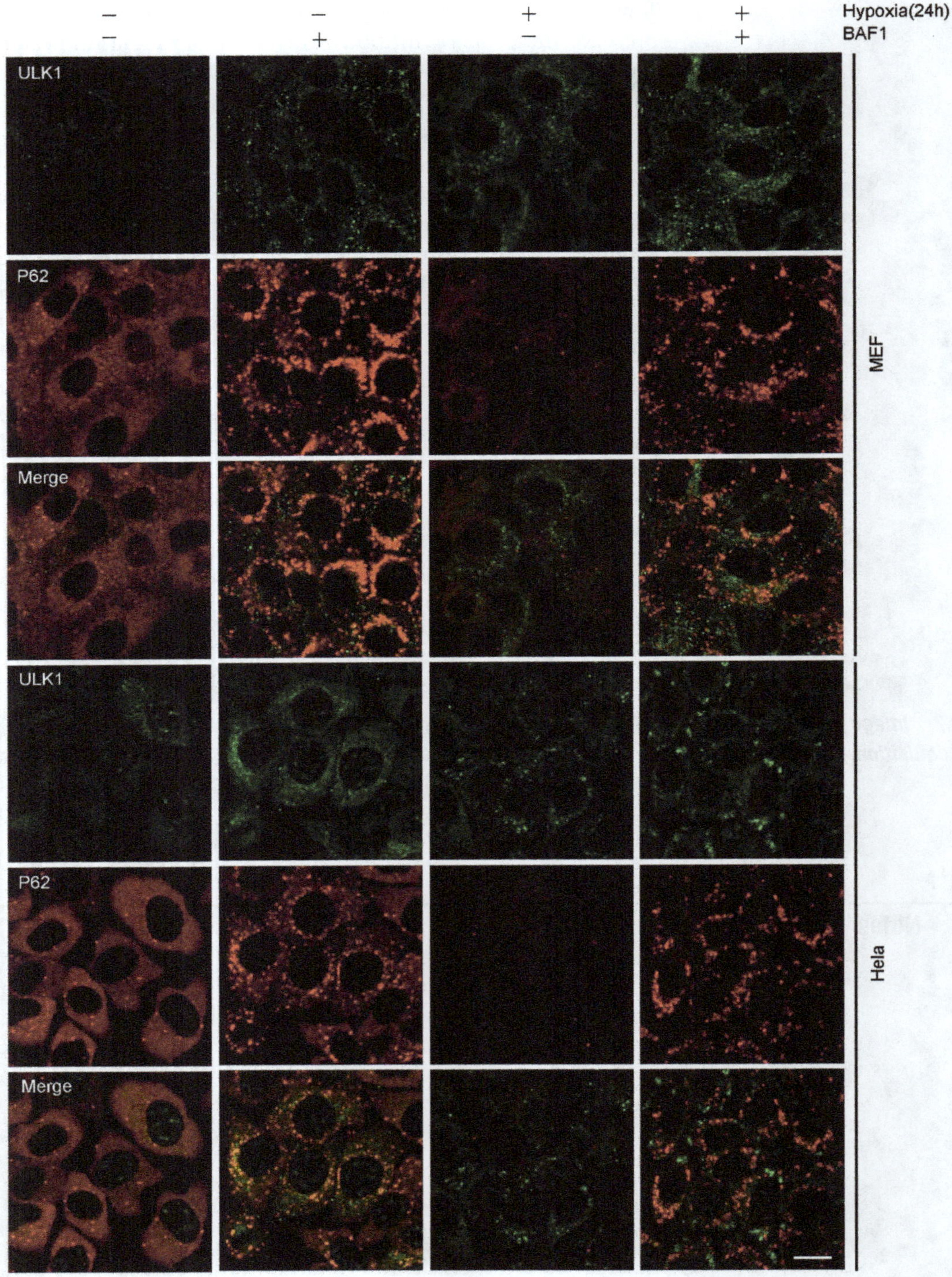

Fig. 1 Immunocytochemical detection of ULK1 and P62 on 4% PFA fixed normal MEF/Hela and hypoxia treated MEF/Hela for 24 h, with ULK1 (A7481, green) and P62 (ab56416, red). The second-step reagents were alexa fluor 488-labeled donkey anti-rabbit IgG antibodies (Invitrogen, A21206) and alexa fluor 555-labeled donkey anti-mouse IgG antibodies (Invitrogen, A31570). Scale bar, 20 μm

4. A small amount of triton X-100 used here helps the antibody to enter into the cells.
5. Here, we will introduce some commonly used antibodies. Anti-ULK1 (Sigma-Aldrich, A7481) is suggested to be used at 1:100 dilution for MEFs and HeLa cells (Fig. 1). Anti-SQSTM1/P62 (Abcam, ab56416) is suitable for MEFs and HeLa cells by using at a concentration of 10 μg/mL (equivalent to 1:100 dilution) (Fig. 1). Different cell lines should select LC3 antibodies of different brands, for example, anti-LC3B (Sigma, L7543) is used at 1:100 dilution for MEFs (Fig. 2) and HeLa cells (Fig. 3); anti-LC3 (MBL, M152-3) is propitious to MEFs for immunofluorescence stain (1:100 dilution) (Fig. 4); anti-LC3 (NOVUS, NB100-2220) is preferred to HUVEC (1:100 dilution) (Fig. 5); nevertheless, anti-LC3 (MBL, PM036) is proved to be more suitable for immunoelectron microscopy.
6. 1% BSA used here is aimed to further reduce the nonspecific binding of secondary antibody.
7. Alexa fluor 555-labeled donkey anti-mouse IgG antibodies (Invitrogen, A31570) is diluted at 1:400; alexa fluor 488-labeled donkey anti-rabbit IgG antibodies (Invitrogen, A21206) is diluted at 1:300.

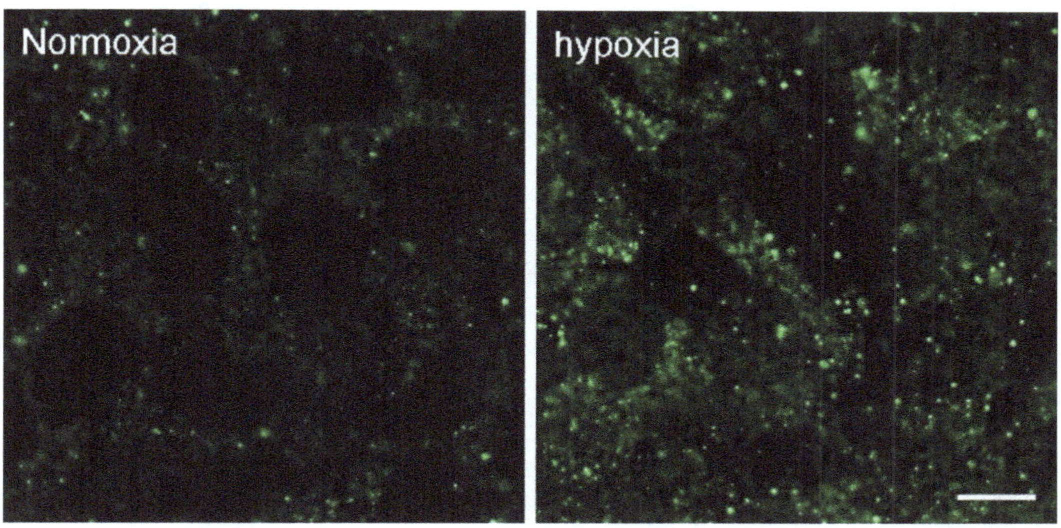

Fig. 2 Immunocytochemical detection of LC3 on 4% PFA fixed normal MEF (left) and hypoxic MEF (right) for 24 h with L7543 (sigma). The second-step reagent was alexa fluor 488-labeled donkey anti-rabbit IgG antibodies (Invitrogen, A21206). Scale bar, 20 μm

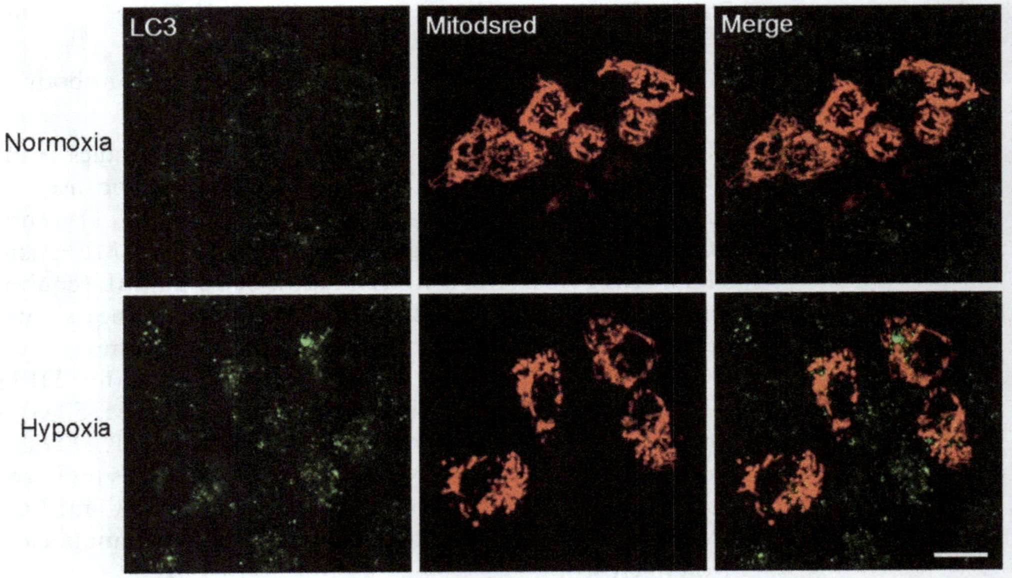

Fig. 3 Confocal immunofluorescent analysis of Hela cells. Cells were transfected with 2 μg mitoDsRed for 24 h, followed by the treatment of normoxia or hypoxia for another 24 h. LC3 (L7543, green), mitoDsRed (red). The second-step reagent was alexa fluor 488-labeled donkey anti-rabbit IgG antibodies (Invitrogen, A21206). Scale bar, 20 μm

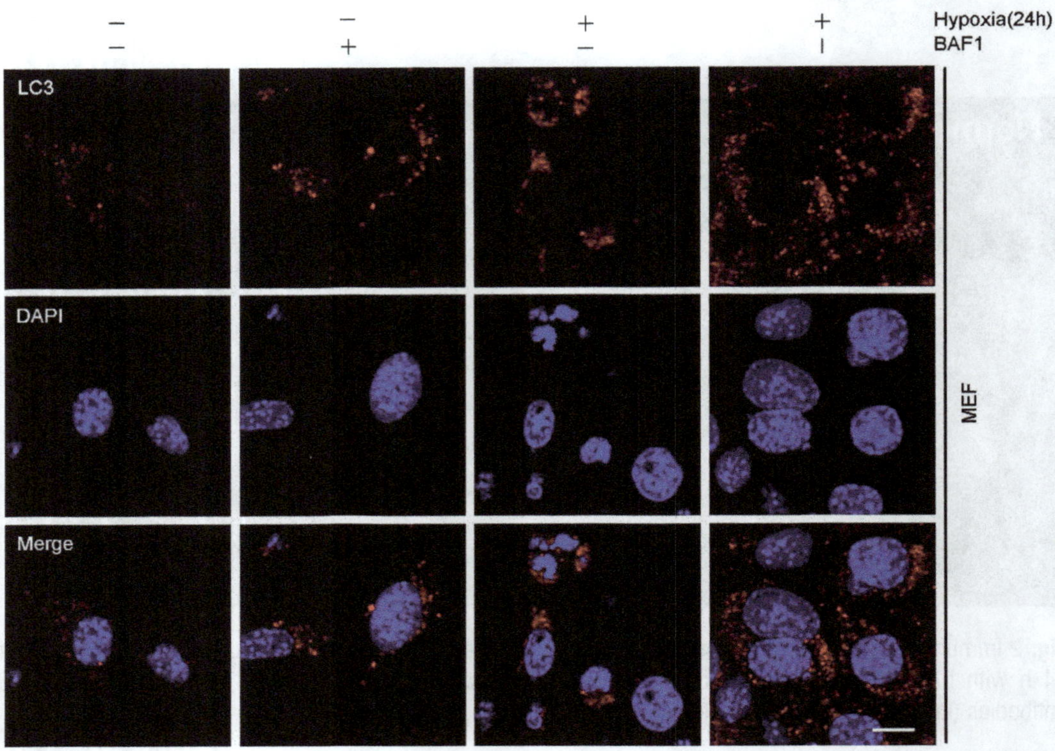

Fig. 4 MEFs were cultured under normoxic or hypoxic condition for 24 h. Cells were stained using anti-LC3 (MBL, M152-3). The second-step reagent was alexa fluor 555-labeled donkey anti-mouse IgG antibodies (Invitrogen, A31570). DAPI was redyed (blue). Scale bar, 20 μm

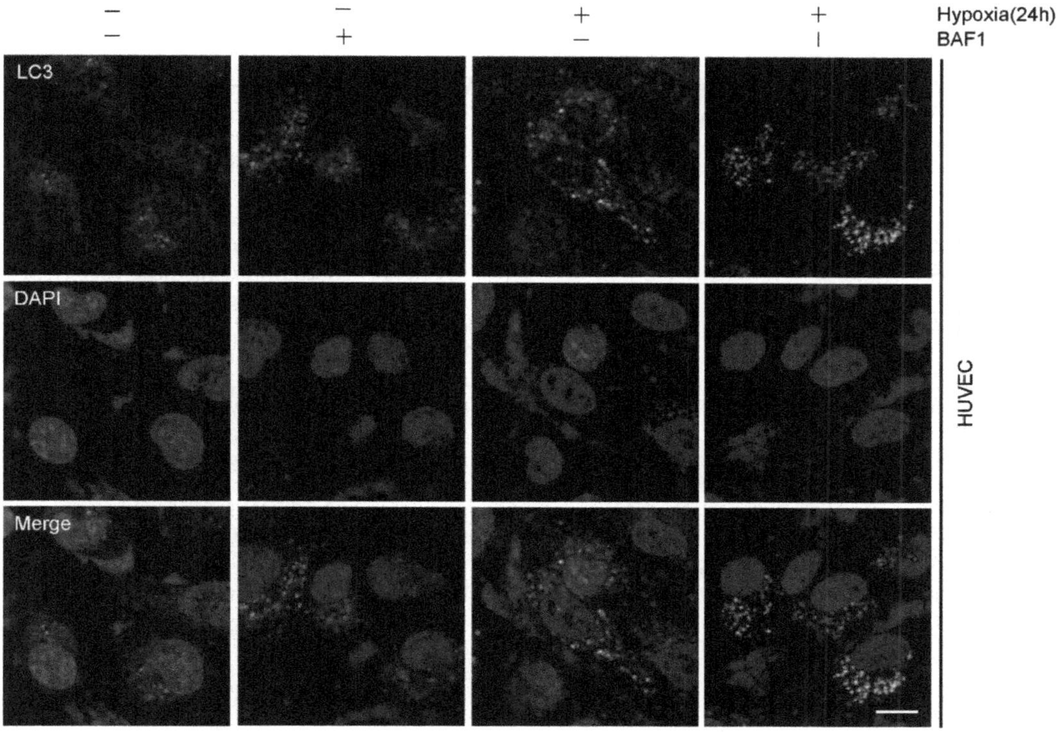

Fig. 5 HUVEC were cultured under normoxic or hypoxic condition for 24 h. Confocal imaging of HUVEC was collected by staining LC3 antibody (green, NOVUS, NB100-2220). The second-step reagent was alexa fluor 488-labeled donkey anti-rabbit IgG antibodies (Invitrogen, A21206). DAPI was redyed (blue). Scale bar, 20 μm

Acknowledgements

This work was supported by NSFC (No. 317781531, No. 91754115), by the Science and Technology Planning Project, Guangdong, China (No. 2017B090901051, No. 2016A020215152), by Research Fund of Guangzhou Medical University (B17017001006), and by Research Fund of Guangdong Medical University (M2014024, M2015001).

References

1. Orhon I, Reggiori F (2017) Assays to monitor autophagy progression in cell cultures. Cell 6 (3). pii: E20). https://doi.org/10.3390/cells6030020
2. Call JA, Wilson RJ, Laker RC, Zhang M, Kundu M, Yan Z (2017) Ulk1-mediated autophagy plays an essential role in mitochondrial remodeling and functional regeneration of skeletal muscle. Am J Physiol Cell Physiol 312:C724–C732
3. Sbrana FV, Cortini M, Avnet S, Perut F, Columbaro M, De Milito A, Baldini N (2016) The role of autophagy in the maintenance of stemness and differentiation of mesenchymal stem cells. Stem Cell Rev 12:621–633

4. Dower CM, Bhat N, Wang EW, Wang HG (2017) Selective reversible inhibition of autophagy in hypoxic breast cancer cells promotes pulmonary metastasis. Cancer Res 77:646–657
5. Guo JY, White E (2016) Autophagy, metabolism, and cancer. Cold Spring Harb Symp Quant Biol 81:73–78
6. Lei Y, Zhang D, Yu J, Dong H, Zhang J, Yang S (2017) Targeting autophagy in cancer stem cells as an anticancer therapy. Cancer Lett 393:33–39
7. Menzies FM, Fleming A, Caricasole A, Bento CF, Andrews SP, Ashkenazi A, Fullgrabe J, Jackson A, Jimenez Sanchez M, Karabiyik C, Licitra F, Lopez Ramirez A, Pavel M, Puri C, Renna M, Ricketts T, Schlotawa L, Vicinanza M, Won H, Zhu Y, Skidmore J, Rubinsztein DC (2017) Autophagy and neurodegeneration: pathogenic mechanisms and therapeutic opportunities. Neuron 93:1015–1034
8. Plaza-Zabala A, Sierra-Torre V, Sierra A (2017) Autophagy and microglia: novel partners in neurodegeneration and aging. Int J Mol Sci 18. pii: E598. https://doi.org/10.3390/ijms18030598
9. Suh HW, Kim JK, Kim TS, Jo EK (2017) New insights into vitamin D and autophagy in inflammatory bowel diseases. Curr Med Chem 24(9):898–910
10. Zhong Z, Sanchez-Lopez E, Karin M (2016) Autophagy, NLRP3 inflammasome and auto-inflammatory/immune diseases. Clin Exp Rheumatol 34:12–16
11. Polus A, Bociaga-Jasik M, Czech U, Goralska J, Cialowicz U, Chojnacka M, Polus M, Jurowski K, Dembinska-Kiec A (2017) The human immunodeficiency virus (HIV1) protease inhibitor sanquinavir activates autophagy and removes lipids deposited in lipid droplets. J Physiol Pharmacol 68:283–293
12. Feizi N, Mehrbod P, Romani B, Soleimanjahi H, Bamdad T, Feizi A, Jazaeri EO, Targhi HS, Saleh M, Jamali A, Fotouhi F, Nargesabad RN, Abdoli A (2017) Autophagy induction regulates influenza virus replication in a time-dependent manner. J Med Microbiol 66:536–541
13. Tian W, Li W, Chen Y, Yan Z, Huang X, Zhuang H, Zhong W, Chen Y, Wu W, Lin C, Chen H, Hou X, Zhang L, Sui S, Zhao B, Hu Z, Li L, Feng D (2015) Phosphorylation of ULK1 by AMPK regulates translocation of ULK1 to mitochondria and mitophagy. FEBS Lett 589:1847–1854
14. Dunlop EA, Tee AR (2013) The kinase triad, AMPK, mTORC1 and ULK1, maintains energy and nutrient homoeostasis. Biochem Soc Trans 41:939–943
15. Kim J, Guan KL (2011) Regulation of the autophagy initiating kinase ULK1 by nutrients: roles of mTORC1 and AMPK. Cell Cycle 10:1337–1338
16. Hosokawa N, Hara T, Kaizuka T, Kishi C, Takamura A, Miura Y, Iemura S, Natsume T, Takehana K, Yamada N, Guan JL, Oshiro N, Mizushima N (2009) Nutrient-dependent mTORC1 association with the ULK1-Atg13-FIP200 complex required for autophagy. Mol Biol Cell 20:1981–1991
17. Park JM, Jung CH, Seo M, Otto NM, Grunwald D, Kim KH, Moriarity B, Kim YM, Starker C, Nho RS, Voytas D, Kim DH (2016) The ULK1 complex mediates MTORC1 signaling to the autophagy initiation machinery via binding and phosphorylating ATG14. Autophagy 12:547–564
18. Lin X, Li S, Zhao Y, Ma X, Zhang K, He X, Wang Z (2013) Interaction domains of p62: a bridge between p62 and selective autophagy. DNA Cell Biol 32:220–227
19. Schimmack G, Schorpp K, Kutzner K, Gehring T, Brenke JK, Hadian K, Krappmann D (2017) YOD1/TRAF6 association balances p62-dependent IL-1 signaling to NF-kappaB. Elife 6. pii: e22416. https://doi.org/10.7554/eLife.22416
20. Yang S, Qiang L, Sample A, Shah P, He YY (2017) NF-kappaB signaling activation induced by chloroquine requires autophagosome, p62 protein, and c-Jun N-terminal kinase (JNK) Signaling and promotes tumor cell resistance. J Biol Chem 292:3379–3388
21. Wang BF, Cao PP, Wang ZC, Li ZY, Wang ZZ, Ma J, Liao B, Deng YK, Long XB, Xu K, Wang H, Wang H, Zeng M, Lu X, Liu Z (2017) Interferon-gamma-induced insufficient autophagy contributes to p62-dependent apoptosis of epithelial cells in chronic rhinosinusitis with nasal polyps. Allergy 72(9):1384–1397
22. Xiao B, Deng X, Lim GGY, Zhou W, Saw WT, Zhou ZD, Lim KL, Tan EK (2017) p62-mediated mitochondrial clustering attenuates apoptosis induced by mitochondrial depolarization. Biochim Biophys Acta 1864:1308–1317
23. Ito S, Kimura S, Warabi E, Kawachi Y, Yamatoji M, Uchida F, Ishibashi-Kanno N, Yamagata K, Hasegawa S, Shoda J, Tabuchi K, Sakai S, Bukawa H, Sekido M, Yanagawa T (2017) Corrigendum to "p62 modulates the intrinsic signaling of UVB-induced apoptosis" [Journal of Dermatological Science 83 (2016) 226-233]. J Dermatol Sci 86:259–260

24. Lam HC, Baglini CV, Lope AL, Parkhitko AA, Liu HJ, Alesi N, Malinowska IA, Ebrahimi-Fakhari D, Saffari A, Yu JJ, Pereira A, Khabibullin D, Ogorek B, Nijmeh J, Kavanagh T, Handen A, Chan SY, Asara JM, Oldham WM, Diaz-Meco MT, Moscat J, Sahin M, Priolo C, Henske EP (2017) p62/SQSTM1 cooperates with hyperactive mTORC1 to regulate glutathione production, maintain mitochondrial integrity, and promote tumorigenesis. Cancer Res 77:3255–3267
25. Conway O, Kirkin V (2017) Love laughs at locksmiths: ubiquitylation of p62 unlocks its autophagy receptor potential. Cell Res 27:595–597
26. Feng Y, Klionsky DJ (2017) Autophagy regulates DNA repair through SQSTM1/p62. Autophagy 13:995–996
27. Duran A, Amanchy R, Linares JF, Joshi J, Abu-Baker S, Porollo A, Hansen M, Moscat J, Diaz-Meco MT (2011) p62 is a key regulator of nutrient sensing in the mTORC1 pathway. Mol Cell 44:134–146
28. Lee YK, Lee JA (2016) Role of the mammalian ATG8/LC3 family in autophagy: differential and compensatory roles in the spatiotemporal regulation of autophagy. BMB Rep 49:424–430
29. Nath S, Dancourt J, Shteyn V, Puente G, Fong WM, Nag S, Bewersdorf J, Yamamoto A, Antonny B, Melia TJ (2014) Lipidation of the LC3/GABARAP family of autophagy proteins relies on a membrane-curvature-sensing domain in Atg3. Nat Cell Biol 16:415–424
30. Metlagel Z, Otomo C, Ohashi K, Takaesu G, Otomo T (2014) Structural insights into E2-E3 interaction for LC3 lipidation. Autophagy 10:522–523
31. Otomo C, Metlagel Z, Takaesu G, Otomo T (2013) Structure of the human ATG12–ATG5 conjugate required for LC3 lipidation in autophagy. Nat Struct Mol Biol 20:59–66
32. Wilkinson DS, Jariwala JS, Anderson E, Mitra K, Meisenhelder J, Chang JT, Ideker T, Hunter T, Nizet V, Dillin A, Hansen M (2015) Phosphorylation of LC3 by the Hippo kinases STK3/STK4 is essential for autophagy. Mol Cell 57:55–68
33. Levitzki A (1985) Reconstitution of membrane receptor systems. Biochim Biophys Acta 822:127–153
34. Tanford C, Reynolds JA (1976) Characterization of membrane proteins in detergent solutions. Biochim Biophys Acta 457:133–170

Methods in Molecular Biology (2018) 1854: 187–195
DOI 10.1007/7651_2018_125

Published online: 23 February 2018

Identification of Novel Autophagy Inhibitors via Cell-Based High-Content Screening

Georgios Konstantinidis, Sonja Sievers, and Yao-Wen Wu

Abstract

Autophagy is a fundamental cellular catabolic pathway mediating the recycling of cellular components. Autophagy has been implicated in pathogenesis of diverse diseases such as neurodegeneration and cancer. Due to the therapeutic potential, the autophagy-modulating agents have profoundly enriched the spectrum of tools used to investigate autophagy. However, many of these compounds have additional off-target effects that may confound elucidation of autophagy in certain contexts. There remains high demand for highly specific and novel chemotypes that can be used to study the regulation mechanism of autophagy and contribute novel pharmacophores for therapeutic purposes. Here, we describe a cell-based quantitative high-content screening (HCS) for autophagy inhibitors using a human breast adenocarcinoma MCF7 cell line stably expressing EGFP-LC3, a bona fide marker of autophagy.

Keywords Autophagy, Autophagy inhibitors, Cell-based screening, High-content screening

1 Introduction

Autophagy (Macroautophagy) is an evolutionarily conserved and genetically programmed intracellular degradation pathway [1]. Cytoplasmic proteins, organelles, and certain pathogens are targeted to the lysosome for degradation in a specific or nonspecific manner. Autophagy continuously operates at basal levels to remove cytoplasmic materials in a nonselective manner to maintain cellular homeostasis. In addition, organelles such as damaged mitochondria or invading pathogens can be removed by selective autophagic processes called mitophagy or xenophagy, respectively. Stress conditions such as nutrient deprivation, hypoxia, reactive oxygen species, DNA damage, and protein aggregates increase autophagic activity [2]. Autophagy plays a critical role in several processes such as development and differentiation [3]. Malfunction of autophagy has been associated with diverse human diseases, including cancer, neurodegeneration, cardiac hypertrophy, and pathogen infection [4].

The autophagy process starts with the formation of a unique double-membrane structure, known as the isolation membrane or phagophore and sequesters autophagic cargos. The growth and expansion of the phagophore was shown to require membranous supply from other compartments such as the Golgi complex, recycling endosomes, and the plasma membrane [5]. Closure of the expanded phagophore leads to the formation of an autophagosome, which fuses with a lysosome to form an autolysosome. The inner membrane of the autophagosome and the engulfed materials within the autophagosome are then degraded by lysosomal hydrolases [6].

Autophagy requires over 30 autophagy-related (Atg) proteins that operate in a concerted hierarchy to drive autophagosome formation [7–9]. Except for the Atg8 family of proteins, other Atg proteins transiently associate with autophagosomal membranes and do not become a part of the autophagosome. Microtubule-associated protein 1 light chain 3 (LC3), the mammalian homolog of Atg8, is an ubiquitin-like protein important for autophagosome formation [10–12]. LC3-I, the cytosolic form of LC3 protein, undergoes ubiquitin-like conjugation with phosphatidylethanolamine (PE) to form LC3-II, which localizes at the autophagosomal membrane [13]. Lipidated LC3 localizes on autophagosomal membranes during the entire lifespan of autophagosome from the initiation of isolation membrane to the autophagosome–lysosome fusion. Therefore, GFP-LC3 has been widely used as a marker to monitor autophagy [11, 12, 14]. Induction of autophagy by starvation or rapamycin leads to an increased number of GFP-LC3 punctate structures representing autophagosomes [15]. Accumulation of GFP-LC3 puncta does not necessarily correlate with the autophagy induction that leads to cargo degradation (autophagic flux) [11, 12, 14]. Impairment of lysosomal function by neutralization of lysosomal pH, for example, via chloroquine or bafilomycin A1 treatment, leads to accumulated GFP-LC3 puncta due to the block of autophagosome–lysosome fusion [11, 12, 14]. Hence, in order to confirm modulation of autophagy flux, additional assays are required, for example, GFP-LC3 puncta quantification in the presence and absence of a lysosomal inhibitor [11, 12] or by monitoring the degradation of the well-characterized autophagic substrate SQSTM1/p62 [16].

Several small-molecule modulators have been identified that target different proteins involved in autophagy. Rapamycin and Torin 1 have been found to inhibit mTOR complex, a negative regulator of autophagy, leading to autophagy induction [17–19]. 3-methyladenine (3MA), wortmannin, and LY294002 inhibit phosphoinositide 3-kinase (PI3K), leading to inhibition of autophagy [20, 21]. Inhibitors of lysosomal acidification such as vacuolar H^+-ATPase (V-ATPase) inhibitor bafilomycin A1 and chloroquine prevent fusion of autophagosome with lysosome [22–24]. Pepstatin A and E-64d inhibit the activity of cathepsins, lysosomal proteases, which participate in the degradation of autophagic cargos. Cell-based screens have led to identification of a number of autophagy modulators [25–27]. Identification of novel autophagy-modulating agents followed by target identification offers tremendous opportunities to discover novel chemotypes and cellular pathways involved in autophagy regulation, providing new insights into this process and opening new avenues for therapeutic intervention [28, 29]. Here, we describe a quantitative cell-based high-content screening (HCS) assay for autophagy inhibitors.

We generated a stable MCF7 cell line expressing EGFP-LC3. Autophagy was induced by nutrient starvation using Earle's balanced salt solution (EBSS), or pharmacologically by inhibition of mTOR using rapamycin. To identify inhibitors for autophagy flux, bafilomycin A1 or chloroquine was included. The assay was validated using the known autophagy inhibitor wortmannin. EGFP-LC3 structures were visualized and quantified as formation of puncta by means of automated microscopy and image analysis (Fig. 1a). In order to assess the ability of the assay to identify active compounds from a large chemical library accurately, a screening window coefficient (Z' factor) was calculated to warrant the quality of the screening (Fig. 1b). To further validate the response of autophagy flux in the cells, EGFP-LC3-II and p62 levels were monitored via western blot under different conditions (Fig. 1c). The assay was used to screen an in-house library of ca. 250,000 compounds at 10 μM. The hits were subjected to dose-dependent assay for determination of IC_{50}s and to further verification by western blot assay.

2 Materials

2.1 Cell Culture, Reagents, and Facilities

1. Media: MEM Eagle, Earle's salts, L-glutamine, sodium bicarbonate (Sigma, M4655); Imaging MEM, no glutamine, no phenol red (Thermo Fisher Scientific, 51200046); OptiMEM, Reduced Serum Medium (Thermo Fisher Scientific, 31985062); EBSS, with sodium bicarbonate, without phenol red (Sigma, E3024); FBS (Thermo Fisher Scientific, 10500064); MEM Non-essential Amino Acid Solution (Sigma, M7145); Sodium pyruvate solution (Sigma, S8636); DPBS (Sigma, D8537); Trypsin-EDTA solution (Sigma, T4049); and Insulin solution human (Sigma, I9278)
2. X-tremeGENE HD DNA Transfection Reagent (Roche, 06 366 236 001)
3. Chemical reagents: DMSO (PanReac AppliChem, A3672); G418 (Serva, 49418); rapamycin (AdipoGen, AG-CN2-0025-M001), wortmannin (Calbiochem, 681676); bafilomycin A1 (BioViotica, BVT-0252-M001); and chloroquine (BioVision, 1825-100)

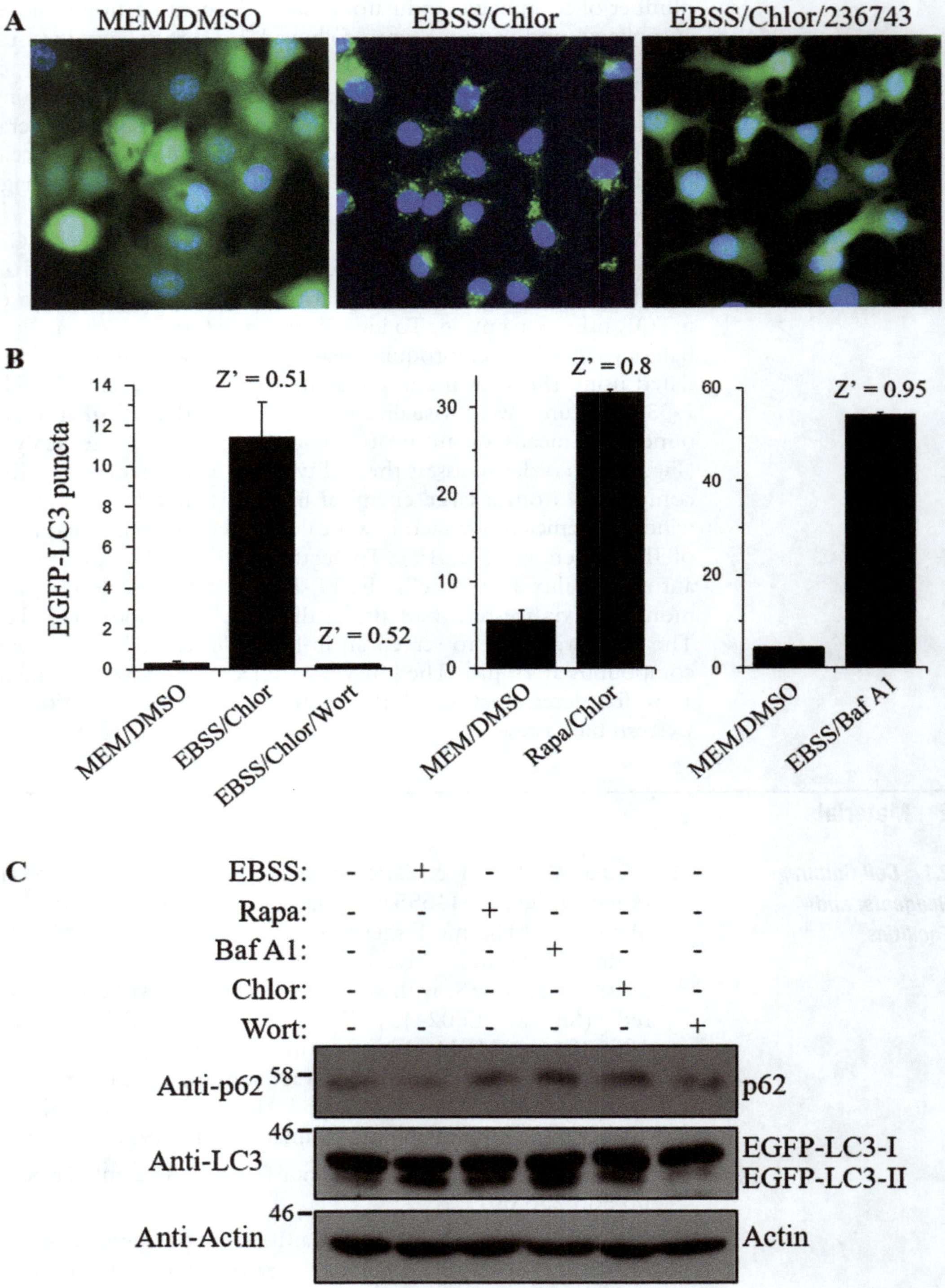

Fig. 1 Phenotypic cellular assay for autophagy inhibitors using the autophagy marker EGFP-LC3. (**a**) Automated confocal imaging of cells. (**b**) Quantification of EGFP-LC3 puncta. (**c**) Western blot assay for autophagy

4. Antibodies: anti-p62 (MBL International Corporation, PM045); anti-LC3 (Cell Signaling, 2775); and anti-Actin (Chemicon, MAB1501)
5. Instrumentation: cell culture microscope (Primo Vert, Zeiss); inverted confocal microscope (Leica TCS SP5 AOBS equipped with a 63×/1.4 HCX Plan Apo oil immersion lens and a temperature-controlled hood at 37 °C and 5% CO_2); Mini-PROTEAN® and Mini Trans-Blot cell (BIO-RAD); and FACS facility (Aria Fusion Flow Cytometry System)

2.2 High-Content Screening

1. 384-well plates black, clear-bottomed
2. Automated dispenser: Multidrop Combi (Thermo)
3. Acoustic dispensing machine: Echo 520 dispenser (Labcyte Inc.)
4. Automated cell washer: Elx405 (Biotek)
5. Dulbecco's phosphate-buffered saline (PBS): 137 mM NaCl, 2.7 mM KCl, 10 mM $Na_2HPO_4 \cdot 2\ H_2O$, 2 mM KH_2PO_4 in ultrapure water H_2O, adjust pH to 7.2–7.4 and autoclave
6. Assay medium: Eagle's MEM, 10% FBS, 1% L-glutamine, 1% sodium pyruvate, 1% NEAA, and 0.01 mg/mL human insulin
7. EBSS
8. Chloroquine (100 mM in H_2O)
9. Starvation medium: EBSS, 50 μM chloroquine
10. Formaldehyde (37%)
11. Hoechst 33342 (1 mg/mL in H_2O)
12. Fix/stain solution: dilute formaldehyde (37% stock solution) to 9.25% with PBS freshly at the screening day and add 0.002 mg/mL Hoechst (1:500)
13. Automated microscope: ImageXpress Micro XL (Molecular Devices) with filter sets for DAPI and FITC
14. Screening library: numerous providers offer screening collections with different focuses, e.g., collections of bioactive molecules which are suitable for repurposing approaches, natural product-derived collections which tend to give a high hit rate, or large collections of screening compounds with unknown function. The assembly of a high-content ready compound library from these large collections requires cheminformatic resources.

Fig. 1 (continued) flux. MEM: complete medium; DMSO: dimethyl sulfoxide; EBSS: Earle's Balanced Salt Solution (starvation medium); Chlor: 50 μM chloroquine; 236743: one of the inhibitors identified (1 μM); Wort: 500 nM wortmannin; Rapa: 500 nM rapamycin; Baf A1: 50 nM bafilomycin A1

3 Methods

3.1 EGFP-LC3 Stable MCF7 Cell Line Generation

Grow MCF7 cells in complete medium: MEM Eagle supplemented with 10% FBS, 1% nonessential amino acids, 1% sodium pyruvate, and 0.01 mg/mL human insulin. Pre-warm all media in a water bath at 37 °C before use. Perform media washing or exchange in MCF7 cells very carefully (*see* **Note 1**).

1. Clone human microtubule-associated protein LC3 into pEGFP-C1 vector (Clontech). The construct contains kanamycin resistance for selection in bacteria and neomycin (G418) for selection in mammalian cells.
2. In 1.5 mL eppendorf tube containing 600 μL OptiMEM, add 6 μg pEGFP-C1-LC3 plasmid. Mix by pipetting.
3. Rinse 80% confluent MCF7 in a p100 dish with DPBS. Add 1 mL Trypsin-EDTA solution, swirl and put the dish back to incubator for 3 min. In meanwhile, add 4.5 μL transfection reagent to plasmid/OptiMEM solution (*see* **Note 2**) and incubate at room temperature for 15 min.
4. Count trypsinized cells and seed 1.8×10^6 cells in a new p100 dish.
5. Add transfection mixture dropwise to the dish. Swirl and put the dish into the incubator.
6. Next day, replace medium with new, supplemented with 400 μg/mL G418. From now on, change medium every 2 days for 3–4 weeks.
7. Perform single-cell Fluorescence-Activated Cell Sorting (FACS) for EGFP (*see* **Note 3**) in 96-well plates. When fluorescent colonies are grown, split into 6-well plates and later on p100 plates. Keep 10% DMSO frozen stocks.
8. Characterize autophagy-related responses of different stable cell lines derived from single cells that homogenously express EGFP-LC3 marker (*see* **Note 4**). Maintain stable cell line of interest in complete medium supplemented with 200 μg/mL G418.

3.2 High-Content Screening

1. Seed stable EGFP-LC3 MCF7 cells in 25 μL assay medium (4000 cells per well) in a black, clear-bottomed 384-well plate and incubate at 37 °C and 5% CO_2 overnight (*see* **Notes 5** and **6**).
2. Wash plates three times with PBS, e.g., using an automated plate washer with final aspiration (*see* **Note 7**).
3. Add 25 nL screening compounds (10 mM stock), e.g., with an acoustic dispensing machine and add 25 μL starvation medium, e.g., using an automated dispenser.
4. Incubate plates for 3 h at 37 °C and 5% CO_2.

5. Add 25 μL fix/stain solution to the cells to fix the cells with simultaneous nuclear staining for 20 min at room temperature.
6. Wash plates three times with PBS, e.g., using an automated plate washer with no final aspiration.
7. Seal the plates using a self-adhesive aluminium foil.
8. Spin down the plates for 1 min at 50 × *g*.
9. Image the plates at 20× magnification in an automated microscope (4 pictures per well).
10. Analyze images using a granularity algorithm, e.g., of the MetaXpress software (Molecular Devices). Use the total autophagosome area per cell parameter for hit evaluation. Hit compounds should reduce total autophagosome area per cell by 60–70% (*see* **Notes 8** and **9**).
11. Validate screening results using fresh compound stock (rebought from the supplier), e.g., in dose–response mode (*see* **Notes 10** and **11**).

4 Notes

1. For media exchange, hold the plate in a 40–50% angle so that media can accumulate and be aspirated from a certain point adjacent to the dish wall. Apply fresh washing or incubation media slowly at the same spot.
2. Release transfection reagent slowly at the center of the plasmid/OptiMEM solution without mixing, pipetting, or contact to the eppendorf plastic wall.
3. When a massive cell death appears and most of the cells wash off the bottom of the dish, leaving colonies of stable (fluorescent) cells behind, start sorting single cells.
4. To characterize autophagy responses in MCF7 cells stably expressing EGFP-LC3 marker, induce (EBSS starvation or 500 nM rapamycin) or inhibit (500 nM wortmannin) autophagy for 2 h. Use 50 nM bafilomycin A1 or 50 μM chloroquine to block autophagosome–lysosome fusion. Image and quantify EGFP-positive punctate structures that represent autophagosomes. Blot with anti-LC3 and determine EGFP-LC3-II (lipidated form) generation under induced conditions or suppression under inhibition conditions. Additionally, blot with anti-p62 to estimate proper autophagic flux by p62 degradation under autophagy-induced conditions.
5. To obtain consistent assay performance and screening results, it is important to use highly similar cell batches. Carefully calculate the number of cells needed for the entire screening and validation procedure. If possible, grow cells in one batch

(or several batches of the same passage number). Freeze the cells in liquid nitrogen until further use. Thaw cells 3 or 4 days before the screening and grow them for one passage to fully recover from freezing. Then, use the cells for the screening. Discard any remaining cells. Always use a new frozen cell batch.

6. It is also possible to seed 1600 cells per well and use them for screening on day 2 after the seeding.
7. Make sure to adjust the washing height properly in order to leave the cell layer intact.
8. Check for cytotoxic effects of the compounds; for example, manually inspect the images of hit compounds.
9. Check for fluorescence artifacts caused by autofluorescence of compounds; for example, manually inspect the images of hit compounds. Autofluorescence of compounds can be so strong that the image analysis algorithm does not work.
10. Validate the purity and integrity of hit compounds by LC-MS analysis.
11. The assay setup can be slightly modified to check whether the compounds act upstream or downstream of mTOR. Instead of starving the cells by using EBSS medium, autophagy can be induced by addition of an mTOR inhibitor (e.g., 500 nM rapamycin).

Acknowledgements

This work was supported by Max Planck Society, Behrens Weise Stiftung, and European Research Council, ERC (ChemBioAP) to Y.W.W.

References

1. Mizushima N, Levine B, Cuervo AM, Klionsky DJ (2008) Autophagy fights disease through cellular self-digestion. Nature 451 (7182):1069–1075. https://doi.org/10.1038/nature06639
2. Kroemer G, Marino G, Levine B (2010) Autophagy and the integrated stress response. Mol Cell 40(2):280–293. https://doi.org/10.1016/j.molcel.2010.09.023
3. Mizushima N, Levine B (2010) Autophagy in mammalian development and differentiation. Nat Cell Biol 12(9):823–830. https://doi.org/10.1038/ncb0910-823
4. Mizushima N, Komatsu M (2011) Autophagy: renovation of cells and tissues. Cell 147 (4):728–741. https://doi.org/10.1016/j.cell.2011.10.026
5. Lamb CA, Yoshimori T, Tooze SA (2013) The autophagosome: origins unknown, biogenesis complex. Nat Rev Mol Cell Biol 14 (12):759–774. https://doi.org/10.1038/nrm3696
6. Yoshimori T (2004) Autophagy: a regulated bulk degradation process inside cells. Biochem Biophys Res Commun 313(2):453–458
7. Xie Z, Klionsky DJ (2007) Autophagosome formation: core machinery and adaptations. Nat Cell Biol 9(10):1102–1109. https://doi.org/10.1038/ncb1007-1102
8. Mizushima N, Yoshimori T, Ohsumi Y (2011) The role of Atg proteins in autophagosome formation. Annu Rev Cell Dev Biol 27:107–132. https://doi.org/10.1146/annurev-cellbio-092910-154005

9. Feng Y, He D, Yao Z, Klionsky DJ (2014) The machinery of macroautophagy. Cell Res 24 (1):24–41. https://doi.org/10.1038/cr.2013.168
10. Ohsumi Y (2001) Molecular dissection of autophagy: two ubiquitin-like systems. Nat Rev Mol Cell Biol 2(3):211–216. https://doi.org/10.1038/35056522
11. Mizushima N, Yoshimori T, Levine B (2010) Methods in mammalian autophagy research. Cell 140(3):313–326. https://doi.org/10.1016/j.cell.2010.01.028
12. Klionsky DJ et al (2016) Guidelines for the use and interpretation of assays for monitoring autophagy (3rd edition). Autophagy 12 (1):1–222. https://doi.org/10.1080/15548627.2015.1100356
13. Tanida I, Ueno T, Kominami E (2008) LC3 and autophagy. Methods Mol Biol 445:77–88. https://doi.org/10.1007/978-1-59745-157-4_4
14. Ni HM, Bockus A, Wozniak AL, Jones K, Weinman S, Yin XM, Ding WX (2011) Dissecting the dynamic turnover of GFP-LC3 in the autolysosome. Autophagy 7(2):188–204
15. Mizushima N, Yamamoto A, Matsui M, Yoshimori T, Ohsumi Y (2004) In vivo analysis of autophagy in response to nutrient starvation using transgenic mice expressing a fluorescent autophagosome marker. Mol Biol Cell 15 (3):1101–1111. https://doi.org/10.1091/mbc.E03-09-0704
16. Johansen T, Lamark T (2011) Selective autophagy mediated by autophagic adapter proteins. Autophagy 7(3):279–296
17. Blommaart EF, Luiken JJ, Blommaart PJ, van Woerkom GM, Meijer AJ (1995) Phosphorylation of ribosomal protein S6 is inhibitory for autophagy in isolated rat hepatocytes. J Biol Chem 270(5):2320–2326
18. Noda T, Ohsumi Y (1998) Tor, a phosphatidylinositol kinase homologue, controls autophagy in yeast. J Biol Chem 273(7):3963–3966
19. Thoreen CC, Kang SA, Chang JW, Liu Q, Zhang J, Gao Y, Reichling LJ, Sim T, Sabatini DM, Gray NS (2009) An ATP-competitive mammalian target of rapamycin inhibitor reveals rapamycin-resistant functions of mTORC1. J Biol Chem 284(12):8023–8032. https://doi.org/10.1074/jbc.M900301200
20. Seglen PO, Gordon PB (1982) 3-Methyladenine: specific inhibitor of autophagic/lysosomal protein degradation in isolated rat hepatocytes. Proc Natl Acad Sci U S A 79 (6):1889–1892
21. Blommaart EF, Krause U, Schellens JP, Vreeling-Sindelarova H, Meijer AJ (1997) The phosphatidylinositol 3-kinase inhibitors wortmannin and LY294002 inhibit autophagy in isolated rat hepatocytes. Eur J Biochem 243 (1–2):240–246
22. Bowman EJ, Siebers A, Altendorf K (1988) Bafilomycins: a class of inhibitors of membrane ATPases from microorganisms, animal cells, and plant cells. Proc Natl Acad Sci U S A 85 (21):7972–7976
23. Yamamoto A, Tagawa Y, Yoshimori T, Moriyama Y, Masaki R, Tashiro Y (1998) Bafilomycin A1 prevents maturation of autophagic vacuoles by inhibiting fusion between autophagosomes and lysosomes in rat hepatoma cell line, H-4-II-E cells. Cell Struct Funct 23 (1):33–42
24. Suzuki T, Nakagawa M, Yoshikawa A, Sasagawa N, Yoshimori T, Ohsumi Y, Nishino I, Ishiura S, Nonaka I (2002) The first molecular evidence that autophagy relates rimmed vacuole formation in chloroquine myopathy. J Biochem 131(5):647–651
25. Fleming A, Noda T, Yoshimori T, Rubinsztein DC (2011) Chemical modulators of autophagy as biological probes and potential therapeutics. Nat Chem Biol 7(1):9–17. https://doi.org/10.1038/nchembio.500
26. Baek KH, Park J, Shin I (2012) Autophagy-regulating small molecules and their therapeutic applications. Chem Soc Rev 41 (8):3245–3263. https://doi.org/10.1039/c2cs15328a
27. Rubinsztein DC, Codogno P, Levine B (2012) Autophagy modulation as a potential therapeutic target for diverse diseases. Nat Rev Drug Discov 11(9):709–730. https://doi.org/10.1038/nrd3802
28. Laraia L, Ohsawa K, Konstantinidis G, Robke L, Wu YW, Kumar K, Waldmann H (2017) Discovery of novel cinchona-alkaloid-inspired oxazatwistane autophagy inhibitors. Angew Chem Int Ed Engl 56(8):2145–2150. https://doi.org/10.1002/anie.201611670
29. Robke L, Laraia L, Carnero Corrales MA, Konstantinidis G, Muroi M, Richters A, Winzker M, Engbring T, Tomassi S, Watanabe N, Osada H, Rauh D, Waldmann H, Wu YW, Engel J (2017) Phenotypic identification of a novel autophagy inhibitor chemotype targeting lipid kinase VPS34. Angew Chem Int Ed Engl 56(28):8153–8157. https://doi.org/10.1002/anie.201703738

Methods in Molecular Biology (2018) 1854: 197–207
DOI 10.1007/7651_2018_158

Published online: 11 May 2018

Porcine Cell-Free System to Study Mammalian Sperm Mitophagy

Won-Hee Song and Peter Sutovsky

Abstract

A cell-free system using oocyte extracts is a valuable tool to study early events of animal fertilization and examine protein-protein interactions difficult to observe in whole cells. The process of postfertilization sperm mitophagy assures timely elimination of paternal, sperm-contributed mitochondria carrying potentially corrupted mitochondrial DNA (mtDNA). Cell-free systems would be especially advantageous for studying postfertilization sperm mitophagy as large amounts of oocyte extracts can be incubated with hundreds to thousands of spermatozoa in a single trial, while only one spermatozoon per zygote can be examined by whole-cell approaches. Since sperm mitophagy is species-specific, the abundantly available frog egg extracts commonly used for cell-free systems have to be replaced with isospecific mammalian oocyte extracts, which are difficult to obtain. Here we describe the protocol for a mammalian, porcine cell-free system consisting of permeabilized domestic boar spermatozoa co-incubated with cell extracts from porcine oocytes, suitable for studying the interactions of maternal, oocyte-derived mitophagy factors with paternal, sperm mitochondria.

Keywords Autophagy, Cell-free system, Extract, Immunostaining, Mitochondria, Oocyte, Porcine, Sperm

1 Introduction

Cellular extracts prepared from *Xenopus laevis* or mammalian oocytes have been established and used extensively for studying many aspects of cellular function [1–4]. The *Xenopus laevis* extracts completed DNA replication of sperm nuclei, providing a rich source of nuclear proteins. The assembly of nuclear pore complexes observed in fertilized zygotes was also reproduced in *Xenopus* oocyte extracts [3]. Pertinent to the optimization of somatic cell nuclear transfer, nuclei of differentiated cells were reprogrammed to an undifferentiated state by cell-free extracts from *Xenopus laevis* or porcine oocytes [1, 2]. Using semi-cell-free system composed of soluble porcine oocyte zona pellucida (ZP) glycoproteins and purified boar spermatozoa, we demonstrated the degradation of ZP proteins by sperm proteasomes, similar to their action during in vitro fertilization [5]. We recently explored a method for examining the binding of oocyte-derived mitophagy factors to sperm mitochondria in a cell-free system consisting of permeabilized mammalian spermatozoa co-incubated with cell extracts from porcine oocytes. This system is useful for studying postfertilization sperm mitophagy, as mammalian sperm mitophagy is a species-specific mechanism that recognizes and degrades sperm mitochondria by oocyte-derived factors only in intraspecific crosses [6, 7]. Another advantage of using porcine oocyte extracts is that thousands of spermatozoa can be treated and analyzed simultaneously, whereas mitochondria from only one spermatozoon can be observed in a zygote.

To obtain meiotically mature porcine oocytes, cumulus-oocyte complexes (COCs) are aspirated from porcine ovarian follicles sized 3–6 mm and matured in tissue culture medium (TCM) 199 supplemented with gonadotropins [8]. Cumulus cells from matured COCs are then denuded and ZP is removed. The ZP-free, mature MII oocytes are washed in extraction buffer containing energy-generating system and lysed in the same buffer by high-speed centrifugation [1]. During oocyte maturation, the cell cycle is arrested at the metaphase of second meiosis (MII). A source of cell-free oocyte extracts at this stage contains maternal components, which are necessary for embryo development after natural fertilization.

It has been known that the ATP energy-generating system is an important component of chromatin assembly and other cellular events, including protein and organelle turnover by autophagy. The reaction of chromatin assembly in *Xenopus* oocyte extracts was reproducible in the presence of ATP [9, 10]. In our protocol, the ATP energy-generating system that is required for sperm chromatin remodeling to form the paternal pronucleus during fertilization was added to porcine oocyte extracts, mimicking the early postfertilization ooplasm.

Mature mammalian sperm nucleus is stabilized with protamines, the sperm-specific arginine-rich nuclear proteins [11]. During epididymal maturation, sulfhydryl groups of cysteine residues of protamines are oxidized to form disulfide bonds (S-S), resulting in the stabilization of sperm nucleus structure [12, 13]. During normal fertilization, oocyte-derived glutathione reduces S-S in the sperm nucleus. This process promotes sperm nuclear decondensation, subsequent formation of the paternal pronucleus, and disassembly of the sperm tail structures [14, 15].

Mammalian sperm mitochondrial membrane is also stabilized by S-S hardened structure [15, 16]. Such structures can be reduced by treatment with reducing agents such as dithiothreitol (DTT). In human and bull spermatozoa, treatment with DTT induced removal of mitochondrial sheaths and disassembly of the sperm tail structures [15], which, in a zygote, would lead to the formation of zygotic centrosome and sperm aster observed during natural fertilization. In the present protocol, boar spermatozoa prior to incubation with oocyte extracts were primed with treatment of L-α-lysophosphatidylcholine (lysolecithin) that removes sperm plasma membrane and subsequently incubated with DTT to reduce S-S bonds in the internal structures of the sperm head and tail. Such treatments are necessary for mimicking sperm demembranation that occurs during sperm-oocyte interactions and pronuclear formation that occurs during the early zygotic development.

By establishing the porcine cell-free system, we were able to recapitulate the early events of postfertilization sperm mitophagy. The oocyte-derived mitophagy factors, SQSTM1 and VCP, were detected on sperm mitochondria after oocyte extract exposure (Fig. 1). Sperm head expansion mimicking the early stages of paternal pronucleus formation was also reproduced by sperm-oocyte extract co-incubation (Fig. 2). Such patterns of sperm-ooplasm interactions are reminiscent of early stage porcine zygotes. Thus, the porcine and other mammalian cell-free systems could be useful for identifying mitophagy factors active during mammalian fertilization and zygotic development. This basic protocol will be further developed to incorporate protein tracers and modifiers suitable for proteomic identification of ooplasmic factors that interact with sperm structures during sperm-oocyte extract co-incubation and in the zygotes obtained by in vitro fertilization.

2 Materials

All solutions are prepared using ultrapure water. All chemicals are purchased from Sigma (St. Louis, MO) unless other commercial suppliers are specified. Reagents are prepared at room temperature unless specified otherwise. Oocytes and oocyte extracts are never exposed to temperatures above 38 °C.

2.1 Preparation of Porcine Oocyte Extracts

1. Meiotically mature porcine oocytes (*see* **Note 1**).
2. HEPES-buffered Tyrode lactate medium containing 0.01% (w/v) PVA (TL-HEPES- PVA): dissolve 114 mM NaCI, 3.2 mM KCI, 0.34 mM NaH_2PO_4, 10 mM Na lactate, 0.5 mM $MgCI_2$, 10 mM HEPES, 0.2 mM Na pyruvate, 12 mM sorbitol, 2 mM $NaHCO_3$, 2 mM $CaCl_2$, and PVA in water. Adjust pH to 7.3–7.4 and filter through a 0.22 μm sterile cellulose acetate membrane before use and store at 4 °C for 3 weeks.
3. 0.1% (w/v) hyaluronidase solution: dissolve hyaluronidase slowly in TL-HEPES-PVA and filter through a 0.22 μm sterile cellulose acetate membrane.
4. 0.1% (w/v) protease solution: dissolve protease in TL-HEPES-PVA and filter through a 0.22 μm sterile cellulose acetate membrane.

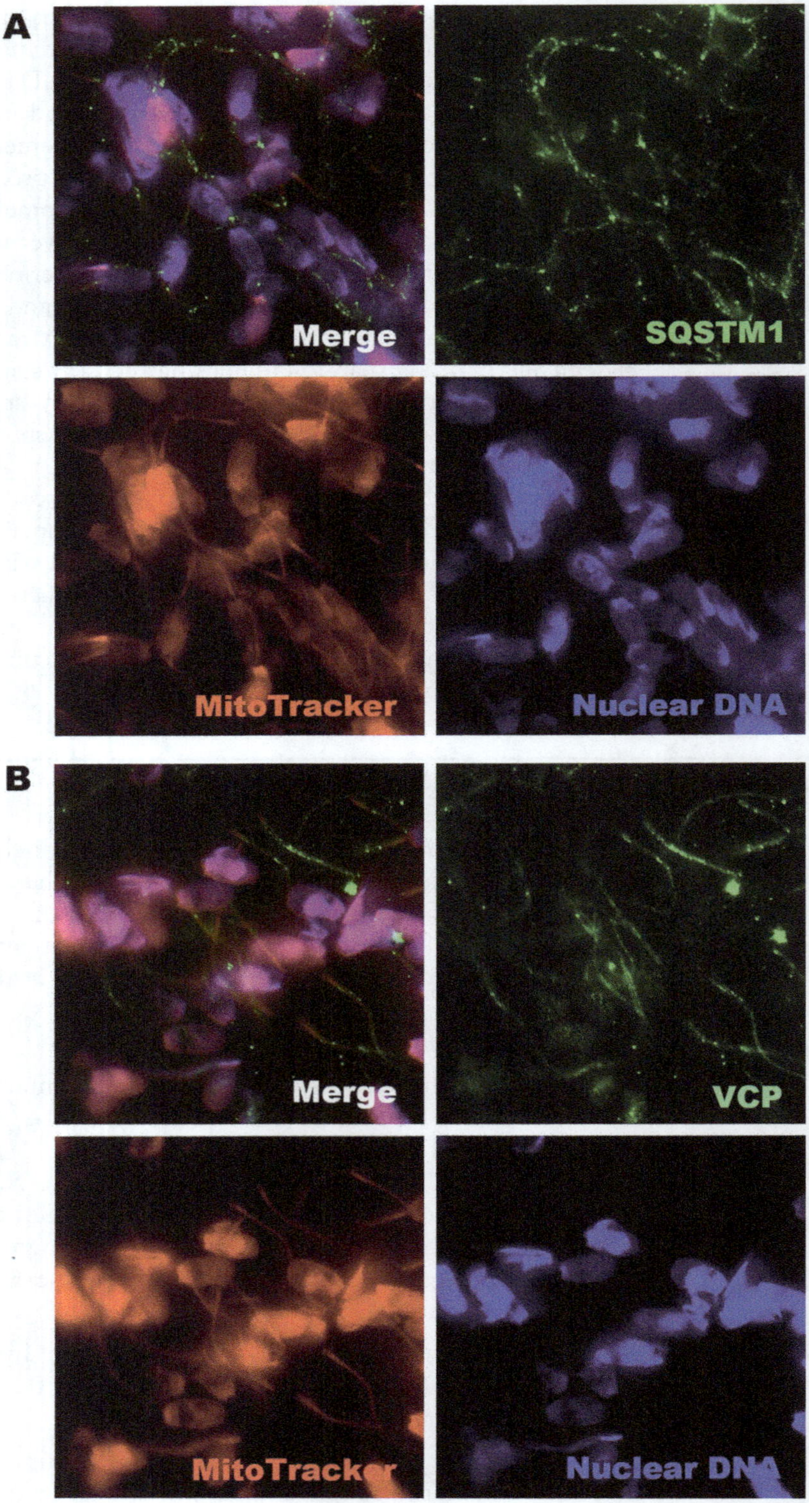

Fig. 1 Immunofluorescence staining of autophagy-/mitophagy-related proteins SQSTM1 and VCP in boar sperm mitochondria exposed to porcine oocyte extracts. Spermatozoa pre-labeled with MitoTracker were

5. Energy-generating reagents: 2 mM ATP, 20 mM phosphocreatine, 20 U/ml creatine kinase, and 2 mM GTP (*see* **Note 2**).
6. Extraction buffer: 50 mM KCl, 5 mM $MgCI_2$, 5 mM ethylene glycol tetraacetic acid (EGTA), 2 mM β-mercaptoethanol, 0.1 mM PMSF, protease inhibitor cocktail (Thermo Fisher Scientific), and 50 mM HEPES. Adjust pH to 7.6 (*see* **Note 3**).
7. 1.5 ml microcentrifuge tube.
8. Single-well 95 mm × 15 mm Petri dish.
9. Liquid nitrogen.
10. Parafilm.
11. String.
12. Vortex (Scientific Industries).
13. Centrifuge (Sorvall Biofuge Fresco).
14. Deep freezer (−80 °C).

2.2 Semen Preparation

1. Boar semen (*see* **Note 4**).
2. Phosphate-buffered saline (PBS): add 137 mM NaCI, 2.68 mM KCl, 1.47 mM KH_2PO_4, and 8.10 mM Na_2HPO_4 in 800 ml. Add water to a total volume of 1 L and adjust pH to 7.4.
3. PBS-polyvinyl alcohol (PVA) 0.1% (w/v) solution: dissolve PVA in PBS, filter through 0.22 μM sterile cellulose acetate membrane, and store at 4 °C.
4. Centrifuge (Thermo Fisher Scientific).
5. Boar semen extender BTS (IMV Technologies).
6. Hemocytometer.
7. MitoTracker® Red CMXRos (Molecular Probes, Thermo Fisher Scientific).

2.3 Sperm Membrane Permeabilization

1. KMT buffer: 100 mM KCl, 2 mM $MgCl_2$, 10 mM Tris–HCl, and 5 mM ethylene glycol tetraacetic acid (EGTA). Adjust pH to 7.0 or 8.2. Filter through a 0.22 μm sterile cellulose acetate membrane before use and store at 4 °C.
2. Permeabilization stock solution: 0.05% (w/v) L-α-lyso phosphatidylcholine (lysolecithin) dilute in KMT (pH 7.0).
3. Dithiothreitol (DTT) stock solution: 10 mM DTT stock solution is prepared with KMT buffer (pH 8.2) (*see* **Note 5**).

Fig. 1 (continued) treated with 0.05% lysolecithin and 10 mM DTT and then co-incubated for 4 h with oocyte extracts. The spermatozoa exposed to extracts were fixed and immunostained with anti-SQSTM1 (**a**) or anti-VCP (**b**) antibodies. Such mitophagy factors were present in the sperm mitochondrial sheath only after co-incubation

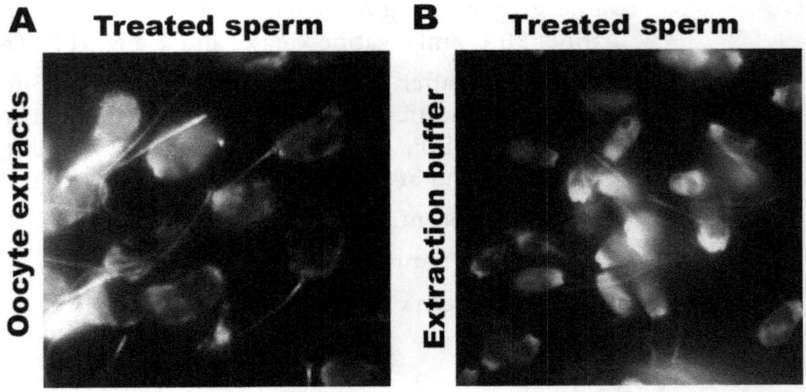

Fig. 2 Boar sperm nucleus expansion after co-incubation with porcine oocyte extracts. Spermatozoa pre-labeled with MitoTracker were treated with 0.05% lysolecithin and 10 mM DTT and were co-incubated for 24 h with oocyte extracts (**a**) or extraction buffer (**b**) as a control. Sperm heads became enlarged during 24 h of co-incubation with the extracts, whereas treated spermatozoa exposed to extraction buffer did not display sperm head expansion

2.4 Co-incubation of Permeabilized Mammalian Spermatozoa with Porcine Oocyte Extracts

1. Poly-L-lysine-coated 18 mm square coverslips or round coverslips (*see* **Note 6**).
2. 15 ml conical centrifuge tube.
3. KMT buffer.
4. Slide warmer.
5. 38.5 °C incubator.

2.5 Immunostaining

1. PBS 0.1% (w/v) NaN_3: dissolve sodium azide in PBS and store at 4 °C.
2. Fixative: use 2% (v/v) formaldehyde (Polysciences, Inc.) solution dilute in PBS-NaN_3.
3. Permeabilization solution: dissolve 0.1% (v/v) Triton X-100 in PBS-NaN_3 (PBS-TX) and mix slowly to avoid bubbles. Store at 4 °C for up to 1 month.
4. Blocking solution: dissolve 5% (v/v) normal goat serum (NGS) in PBS-TX.
5. Washing solution: dissolve 1% (v/v) NGS in PBS-TX.
6. Single-well 95 mm × 15 mm Petri dishes.
7. Four-well Petri dish.
8. Primary and secondary antibodies (*see* **Note 7**).
9. Nucleic acid stain (DAPI).
10. Aluminum foil.
11. Microcopy glass slides.

12. Antifade mounting medium (Vectashield, Vector Labs).
13. Clear nail polish for sealing slide-mounted coverslips.
14. Epifluorescence microscope.

3 Methods

3.1 Preparation of Porcine Oocyte Extracts

1. Transfer mature oocytes into a 1.5 ml microcentrifuge tube, add 0.1% (w/v) hyaluronidase solution, and vortex for 5 min to denude oocytes or remove cumulus cells.
2. In a Petri dish, prepare a drop of 0.1% (w/v) protease solution and three drops of TL-HEPES-PVA. Transfer cumulus-free oocytes into a drop of the protease solution to remove ZP and then wash three times by transferring through three drops of TL-HEPES-PVA.
3. Transfer the ZP-free, mature oocytes into a new 1.5 ml microcentrifuge tube and store in a deep freezer (−80 °C) (*see* **Note 8**).
4. Add extraction buffer containing the energy-generating agents into the ZP-free, mature oocytes in 1.5 ml microcentrifuge tube, seal the tube with parafilm, and tie a string around it (*see* **Note 9**).
5. Lower the tube on a string into liquid nitrogen tank, and let it freeze for 10–30 s. By pulling the string, carefully remove the tube from the liquid nitrogen to thaw it at room temperature.
6. Repeat **step 5** three times to achieve oocyte homogenization by freeze-thawing cycles, and then keep the tube in ice.
7. Crush the oocytes in the microcentrifuge tube by centrifugation at 16,000 × *g* for 20 min at 4 °C.
8. Collect the supernatant (oocyte extracts) into a new 1.5 ml microcentrifuge tube.
9. Keep the extracts in a deep freezer (−80 °C) until use.

3.2 Sperm Priming for Cell-Free System

1. To prime sperm mitochondrial sheaths for cell-free studies, dilute boar semen stored in BTS to a concentration of 1×10^4 spermatozoa/ml, put 1 ml spermatozoa in a 15 ml centrifuge tube, wash with PBS-PVA twice by centrifugation at 800 × *g* for 5 min, and then discard the supernatant.
2. Add 1 ml PBS-PVA into the sperm pellets and incubate with 400 nM MitoTracker CMXRos or similar, diluted in KMT, in a 37 °C warm incubator for 10 min (*see* **Note 10**).
3. After pre-labeling with MitoTracker, wash spermatozoa twice with PBS-PVA by centrifugation at 800 × *g* for 5 min, and then discard the supernatant.

4. Permeabilize the sperm pellets with 0.05% (w/v) lysolecithin diluted in KMT, pH 7.0 for 10 min in a 37 °C warm incubator, and wash with the KMT twice for 5 min by centrifugation.
5. Incubate the sperm pellets with 10 mM DTT diluted in KMT, pH 8.2 for 20 min in 37 °C warm incubator, wash with KMT twice for 5 min by centrifugation to terminate the reaction, and discard the supernatant.

3.3 Co-incubation of Permeabilized Mammalian Spermatozoa with Porcine Oocyte Extracts

1. Once the spermatozoa are treated with lysolecithin and DTT, add 10 μl porcine oocyte extracts into the sperm pellet placed in a 15 ml centrifuge tube and mix carefully by gently pipetting the mixture in and out (*see* **Note 11**).
2. Co-incubate for 4–24 h in a humid atmosphere at 38.5 °C.
3. At the end of co-incubation, place poly-L-lysine-coated coverslips on a slide warmer; pipette a 300 ml of KMT solution onto them.
4. Transfer 1 μl of the spermatozoa exposed to oocyte extracts into a drop of KMT solution on the coverslip and incubate for 7 min at 38.5 °C on a slide warmer.
5. To prepare samples for immunocytochemistry, transfer the sperm-coated coverslips into 2% electron microscopy grade formaldehyde solution in PBS-NaN_3 to fix spermatozoa exposed to extracts; 4% paraformaldehyde in PBS can be substituted.

3.4 Immunocytochemistry to Visualize Mitophagy Factors in the Cell-Free System

1. Transfer the spermatozoa affixed onto poly-L-lysine-coated coverslips from the fixative to a dish with PBS-NaN_3, and rinse three times with PBS-NaN_3 to remove the fixative.
2. Prepare 0.1% Triton X-100 in PBS-NaN_3 in a Petri dish to permeabilize the spermatozoa. Immerse the sperm-coated coverslips in the Petri dish and incubate in 0.1% Triton X-100 in PBS-NaN_3 for 40 min at room temperature.
3. For blocking non-specific antibody binding, place each sperm-coated coverslip onto a four-well Petri dish; put 100 μl PBS-TX containing 5% NGS onto the coverslips and incubate for 30 min.
4. Dilute the primary antibody to desired concentration in 100 μl PBS-TX containing 1% NGS in 1.5 ml microcentrifuge tube. Pipette the primary antibody onto the sperm-coated coverslip and incubate for 1 h at 37 °C or overnight at 4 °C.
5. Transfer the sperm-coated coverslip in a Petri dish of PBS-TX containing 1% NGS. Wash three times by carefully immersing in the PBS-TX containing 1% NGS.
6. Dilute secondary antibody in PBS-TX containing 1% NGS in 1.5 ml microcentrifuge tube. Add 2.5 μg/ml of nucleic acid stain DAPI to the secondary antibody solution. Pipette the

mixture onto the sperm-coated coverslip and incubate in dark at room temperature for 40 min (*see* **Note 12**).

7. Repeat washing step by immersing in the PBS-TX containing 1% NGS.
8. After washing, carefully remove the solution from the other side of coverslips.
9. Put a small drop (~10 μl) of antifade mounting medium on a glass slide.
10. Carefully place the sperm-coated coverslips side down, on the drop of mounting medium on the slide glass.
11. Seal coverslip edges with clear nail polish.
12. Keep the slides in the dark at 4 °C until observation.

3.5 Microscopy

1. Use an epifluorescence microscope with a digital camera and appropriate filter sets corresponding to the antibodies used during the immunocytochemistry.
2. Place a small drop of immersion oil on the coverslip, and use high-magnification immersion objectives.

4 Notes

1. Collect ZP-free mature oocytes into 1.5 microcentrifuge tubes. Store in a deep freezer (−80 °C). Add 100 μl of extraction buffer to 1000 porcine mature oocytes per tube.
2. Prepare a stock solution of 2000 U/ml creatine kinase dissolved in 0.5% $NaHCO_3$. Dissolve ATP, phosphocreatine, and GTP in water, and then add creatine kinase stock solution into the final solution to get the desired concentration.
3. Prepare 10 mM PMSF stock dissolved in EtOH. Dissolve KCl, $MgCl_2$, EGTA, protease inhibitor cocktail, and HEPES in water. Add PMSF stock to the final solution to get the desired concentration. Mix and adjust pH to 7.6. The extraction buffer is then mixed with energy-generating system. Add beta-mercaptoethanol to the final solution after mixing the energy-generating system.
4. Fresh boar semen should be used. Centrifuge fresh boar semen at 800 × *g* for 10 min to separate spermatozoa from seminal plasma. Dilute semen with BTS extender to a final concentration of 1×10^8 spermatozoa/ml and store in a Styrofoam box at room temperature for up to 5 days.
5. Only freshly made DTT solution prepared and maintained in 37 °C warm buffer for no more than 30 min prior to application should be used and kept protected from light (e.g., aluminum foil).

6. Dissolve 5 mg of poly-L-lysine in 50 ml water, and the poly-L-lysine solution should be kept in a freezer (−20 °C). Place clean 18 mm coverslips on a slide warmer; pipette approximately 300 ml of poly-L-lysine solution onto the surface of coverslips and incubate for 7 min on slide warmer. Remove poly-L-lysine solution carefully with lens paper, dry the coverslips completely, and store the coated slides at room temperature. Detailed protocols of immunocytochemical procedures can be found in Sutovsky [17].
7. To detect sperm mitophagy-associated oocyte proteins, antibodies specific to SQSTM1, GABARAP, LC3, and VCP are used as primary antibodies. Their corresponding fluorescent-labeled secondary antibodies can then be used. Details are in Song et al. [18].
8. The ZP-free, mature oocytes can be collected over an extended period of time and cryobanked at −80 °C in batches of 1000 oocytes/tube for future use.
9. Add 100 μl of extraction buffer to 1000 ZP-free, mature porcine oocytes.
10. It is necessary to use a high-concentration (400 nM) Mito-Tracker (relative to lower concentrations used for sperm labeling prior to in vitro fertilization) to withstand the treatment with lysolecithin and DTT. Treatments can cause spreading of the probe to sperm head and sperm tail principal piece.
11. Ten microliters of oocyte extracts are derived from 1000 porcine oocytes/batch (100 μl extraction buffer).
12. From this step on, always protect the samples from light by covering them with aluminum foil.

References

1. Miyamoto K, Tsukiyama T, Yang Y, Li N, Minami N, Yamada M, Imai H (2009) Cell-free extracts from mammalian oocytes partially induce nuclear reprogramming in somatic cells. Biol Reprod 80(5):935–943. https://doi.org/10.1095/biolreprod.108.073676
2. Miyamoto K, Furusawa T, Ohnuki M, Goel S, Tokunaga T, Minami N, Yamada M, Ohsumi K, Imai H (2007) Reprogramming events of mammalian somatic cells induced by Xenopus laevis egg extracts. Mol Reprod Dev 74(10):1268–1277. https://doi.org/10.1002/mrd.20691
3. Sutovsky P, Simerly C, Hewitson L, Schatten G (1998) Assembly of nuclear pore complexes and annulate lamellae promotes normal pronuclear development in fertilized mammalian oocytes. J Cell Sci 111(Pt 19):2841–2854
4. Blow JJ, Laskey RA (1986) Initiation of DNA replication in nuclei and purified DNA by a cell-free extract of Xenopus eggs. Cell 47(4):577–587
5. Zimmerman SW, Manandhar G, Yi YJ, Gupta SK, Sutovsky M, Odhiambo JF, Powell MD, Miller DJ, Sutovsky P (2011) Sperm proteasomes degrade sperm receptor on the egg zona pellucida during mammalian fertilization. PLoS One 6(2):e17256. https://doi.org/10.1371/journal.pone.0017256
6. Kaneda H, Hayashi J, Takahama S, Taya C, Lindahl KF, Yonekawa H (1995) Elimination of paternal mitochondrial DNA in intraspecific

crosses during early mouse embryogenesis. Proc Natl Acad Sci U S A 92(10):4542–4546
7. Shitara H, Hayashi JI, Takahama S, Kaneda H, Yonekawa H (1998) Maternal inheritance of mouse mtDNA in interspecific hybrids: segregation of the leaked paternal mtDNA followed by the prevention of subsequent paternal leakage. Genetics 148(2):851–857
8. Abeydeera LR, Wang WH, Prather RS, Day BN (1998) Maturation in vitro of pig oocytes in protein-free culture media: fertilization and subsequent embryo development in vitro. Biol Reprod 58(5):1316–1320
9. Glikin GC, Ruberti I, Worcel A (1984) Chromatin assembly in Xenopus oocytes: in vitro studies. Cell 37(1):33–41
10. Ryoji M, Worcel A (1984) Chromatin assembly in Xenopus oocytes: in vivo studies. Cell 37 (1):21–32
11. McLay DW, Clarke HJ (2003) Remodelling the paternal chromatin at fertilization in mammals. Reproduction 125(5):625–633
12. Perreault SD, Wolff RA, Zirkin BR (1984) The role of disulfide bond reduction during mammalian sperm nuclear decondensation in vivo. Dev Biol 101(1):160–167
13. Perreault SD (1992) Chromatin remodeling in mammalian zygotes. Mutat Res 296(1–2):43–55
14. Sutovsky P, Schatten G (1997) Depletion of glutathione during bovine oocyte maturation reversibly blocks the decondensation of the male pronucleus and pronuclear apposition during fertilization. Biol Reprod 56(6):1503–1512
15. Sutovsky P, Tengowski MW, Navara CS, Zoran SS, Schatten G (1997) Mitochondrial sheath movement and detachment in mammalian, but not nonmammalian, sperm induced by disulfide bond reduction. Mol Reprod Dev 47 (1):79–86. https://doi.org/10.1002/(SICI)1098-2795(199705)47:1<79::AID-MRD11>3.0.CO;2-V
16. Bedford JM, Calvin HI (1974) Changes in -S-S- linked structures of the sperm tail during epididymal maturation, with comparative observations in sub-mammalian species. J Exp Zool 187(2):181–204. https://doi.org/10.1002/jez.1401870202
17. Sutovsky P (2004) Visualization of sperm accessory structures in the mammalian spermatids, spermatozoa, and zygotes by immunofluorescence, confocal, and immunoelectron microscopy. Methods Mol Biol 253:59–77. https://doi.org/10.1385/1-59259-744-0:059
18. Song WH, Yi YJ, Sutovsky M, Meyers S, Sutovsky P (2016) Autophagy and ubiquitin-proteasome system contribute to sperm mitophagy after mammalian fertilization. Proc Natl Acad Sci U S A 113(36):E5261–E5270. https://doi.org/10.1073/pnas.1605844113

Methods in Molecular Biology (2018) 1854: 209–222
DOI 10.1007/7651_2018_159

Published online: 01 June 2018

Monitoring and Measuring Mammalian Autophagy

Sujit K. Bhutia, Prakash P. Praharaj, Chandra S. Bhol, Debasna P. Panigrahi, Kewal K. Mahapatra, Srimanta Patra, Sarbari Saha, Durgesh N. Das, Subhadip Mukhopadhyay, Niharika Sinha, Prashanta K. Panda, and Prajna P. Naik

Abstract

Macroautophagy (autophagy) is a conserved lysosomal-based intracellular degradation pathway. Here, we present different methods used for monitoring autophagy at cellular level. The methods involve Atg8/LC3 detection and quantification by Western blot, autophagic flux measurement through Western blot, direct fluorescence microscopy or indirect immunofluorescence, and finally traffic light assay using tf-LC3-II. Monitoring autophagic flux is experimentally challenging but obviously a prerequisite for the proper investigation of the process. These methods are suitable for screening purposes and can be used for measurements in cell lysates as well as in living cells. These assays have proven useful for the identification of genes and small molecules that regulate autophagy in mammalian cells.

Keywords Autophagic flux, LC3-II, p62, TEM, tf-LC3

1 Introduction

Autophagy is an evolutionary conserved quality control mechanism involved in the bulk degradation of both long-lived cytoplasmic proteins and dysfunctional organelles to maintain cellular homeostasis for normal body functioning [1]. Autophagy initiates with the formation of phagophore, which maturates into a double-layered dense autophagosome. Autophagosome then fuses with lysosomes to form autolysosomes, allowing the degradation of cytoplasmic cargo and subsequent transportation from the autolysosome for the reuse by the cell [2, 3]. Transmission electron microscopy (TEM) is the only available technique that can visualize the double-membrane autophagophore structures, which is the hallmark of the autophagy process and shows the structure and component of individual cells in their natural environment [4]. TEM can monitor both selective and nonselective autophagy. For selective autophagy, the cargo that is the specific substrate being targeted for sequestration of bulk cytoplasm is essentially excluded. Accumulation of autophagosomes can be measured by TEM image analysis [5, 6]. During autophagy induction, LC3B is proteolytically cleaved at the C terminus by a cysteine protease Atg4, to generate LC3B-I. Further, Atg3, an E2-like carrier, interacts with LC3-I and promotes lipidation to give rise to LC3-I-phosphatidylethanolamine (PE) conjugate or LC3-II which tightly associates with the autophagosomal membrane. During autophagic process, LC3-II is degraded in autolysosomes which is a widely used marker for monitoring the autophagic process [7, 8]. On SDS-PAGE, LC3-II migrates faster than LC3-I which usually gives bands: LC3-I (apparent mobility, 18 kDa) and LC3-II (apparent mobility, 16 kDa). However, an increased level of LC3-II expression is not always an indication of autophagy induction rather it may represent a blockade in autophagosome maturation. Therefore, it is advised to perform autophagic flux in the presence of lysosomal degradation inhibitors, such as pepstatin A, E64d, bafilomycin A1, chloroquine, and NH_4Cl [6, 9]. The consequent translocation of LC3 from diffused form into puncta has become a standard indicator of autophagosome formation and can be visualized either by antibody staining or by fluorescent microscopy of fusion proteins like tf-LC3. Another often used flux marker is sequestosome/p62; this protein binds members of the Atg8 family and functions as a cargo receptor for proteins destined for the autophagosome and is itself degraded via autophagy [10–12].

2 Materials

Prepare all the solutions in ultrapure water with analytical grade reagents and store them at room temperature (unless specific storage condition mentioned). Follow all the waste disposal regulations carefully while disposing waste materials.

2.1 Transmission Electron Microscopy (TEM)

1. *Phosphate buffer*: Prepare a 0.2 M phosphate buffer consisting of 0.2 M NaH_2PO_4 (sodium dihydrogen orthophosphate) (HIMEDIA #RM1257) and 0.2 M Na_2HPO_4 (disodium hydrogen orthophosphate) (SIGMA #V800397). Store at 4 °C.
2. *Glutaraldehyde (primary fixative)*: (2.5% glutaraldehyde) Mix 50 ml of 50% glutaraldehyde (SIGMA #49629) with 950 ml of sodium cacodylate buffer (SIGMA #97068) or 1% glutaraldehyde/10% formalin in phosphate buffer and paraformaldehyde/glutaraldehyde (2% each in 0.1 M buffer) (*see* **Note 1**).
3. *Osmium tetroxide (secondary fixative)*: [0.5% osmium tetroxide in sodium cacodylate buffer (SIGMA #75632)]. Take 200 ml of sodium cacodylate buffer in a brown reagent bottle and add

osmium (add an ampoule if you cannot get all of the crystals out) and mix thoroughly (container capped and sealed with parafilm) to dissolve. Store in a refrigerator. Alternatively, one can prepare a 4% stock solution in deionized water and store aliquots in the freezer (*see* **Note 2**).

4. *Uranyl acetate contrast (enhancing stain)*: Dissolve 2% uranyl acetate (SIGMA #73943) in 50% ethanol or use 4% uranyl acetate in 40% ethanol. Shake well and allow to settle for overnight before usage. Store in an amber color glass bottle (*see* **Note 3**).
5. *Lead citrate contrast (enhancing stain)*: Take 1.33 g lead nitrate (SIGMA #15326) and 1.76 g sodium citrate (SIGMA #71497) in a 50 ml conical tube and add 30 ml of CO_2-free deionized water. Shake vigorously for 1 min. Continue shaking intermittently for 30 min. Add 5 ml of 1 N sodium hydroxide (NaOH) (HIMEDIA #MB095) to the mixture and swirl. Make the final volume up to 50 ml and filter the solution through a syringe filter. Store at 4 °C (*see* **Note 4**).

2.2 Acidic Content Analysis

1. *Acridine orange hydrochloride solution* (Sigma #A8097-10ML) (*see* **Note 5**).
2. *Phosphate-buffered saline* (PBS 1×) (NaH_2PO_4 0.39 g, Na_2HPO_4 1.45 g, NaCl 9.02 g, pH 7.4).
3. *Dansylcadaverine* (SIGMA #30432).
4. *Equipment* (Olympus IX71 fluorescence microscope, BD ACCURI C6 plus flow cytometer).

2.3 Atg8/LC3, P62, and Other Autophagic Protein Detection and Quantification by Western Blot

1. *Cell lysis buffer* (Cell Signaling Technology #9803S) [50 mM Tris–HCl (pH 7.6), 25 mM NaCl, 0.5% Triton X-100, 2 mM dithiothreitol (DTT)]. Dilute the 10× lysis buffer to make it 1× and add 1× cocktail protease inhibitors (Roche Applied Science), 1 mM phenylmethylsulfonyl fluoride (PMSF), 10 mM sodium fluoride, and 1 mM sodium orthovanadate.
2. *4× Laemmli sample buffer* (BioRad #161-0747) [277.8 Mm Tris–HCl, pH 6.8, 44.4% (V/V) glycerol, 4.4% LDS (lithium dodecyl sulfate), 0.02% bromophenol blue].
3. *Acrylamide/Bis solution (30% w/v)* (29.2:0.8 acrylamide/Bis). Weigh 29.2 g of acrylamide monomer and 0.8 g Bis (cross-linker) and transfer to a 100 ml graduated cylinder containing about 40 ml of water. Mix for about 30 min. Make up to 100 ml with water and filter through a 0.45 μm Corning filter. Store at 4 °C, in an amber color bottle (*see* **Note 6**).
4. *Ammonium persulfate*: 10% solution in water.
5. *SDS-PAGE running buffer*: 0.025 M Tris–HCl, pH 8.3, 0.192 M glycine, 0.1% SDS (*see* **Note 7**).

6. *Western blot transfer buffer* (0.025 M Tris, 0.192 M glycine, and 20% methanol).
7. *Washing buffer*: 1× PBS containing 0.05% Tween-20 (PBST).
8. *Blocking solution* (5% skim milk/BSA in PBST). Add 5 g of skim milk in 100 ml of PBST and mix it. Store at 4 °C.

2.4 GFP-Atg8/LC3 and Colocalization Study by Fluorescence and Confocal Microscopy

1. Plasmid transfection: *GFP-Atg8/LC3 plasmid (addgene #22405)*, Lipofectamine™ 2000 Transfection Reagent (Invitrogen *#11668027*).
2. *Fixing solution*: Methanol (100%) or formalin (10%).
3. *Washing buffer*: PBS containing 0.05% Tween-20 (PBST).
4. *Blocking solution blocking buffer* (1× PBS/5% normal serum/ 0.3% Triton™ X-100). Add 0.5 ml normal serum from the same species as the secondary antibody [e.g., normal goat serum (Cell Signaling and Technology *#5425*) to 9.5 ml 1× PBS] and mix well. Add 30 μl Triton™ X-100.
5. *Antibody dilution buffer* (1× PBS/1% BSA/0.3% Triton™ X-100). Add 30 μl Triton™ X-100 to 10 ml 1× PBS. Mix well then add 0.1 g BSA (Cell Signaling and Technology *#9998*).
6. *Antibodies* (LC3B (1:1000; Novus Biologicals #nb-100-2220), SQSTM1/p62 (1:1000, BD Biosciences #610832), ULKI (1:1000, Cell Signaling and Technology #8054S), Beclin-1 (1:1000, Cell Signaling and Technology #3738), Atg5 (1:1000, Cell Signaling and Technology #2630S), Atg7 (1:1000 Biogenesis-R-161-100), and anti-rabbit and/or anti-mouse antibodies conjugated with Alexa Fluor (1:500; A11001, A11004, A11011, and A11008 from Invitrogen).
7. *Equipments* [Olympus IX71 fluorescent inverted microscope and the cellSens standard software (version 1.6, Olympus Soft Imaging Solutions GmbH, Johann-Krane-Weg, Münster, Germany)].

2.5 Measurement of Autophagic Flux

2.5.1 LC3-II/Atg8–PE Turnover

1. *Cell culture requirements*: Culture media (desired media with 10% FBS), Trypsin-EDTA solution (HIMEDIA #TCL007), antibiotic/antimycotic solution (HIMEDIA #A002A), Dulbecco's phosphate-buffered saline (DPBS) (HIMEDIA #TL1023).
2. *Western requirements*: Cell lysis buffer (Cell Signaling Technology #9803S), 4× Laemmli sample buffer (BioRad #161-0747), nitrocellulose membrane (Pall Corporation #66485), bafilomycin A1 (Sigma Aldrich #B1793), LC3-II antibody (Sigma Aldrich #L7543), p62 antibody (BD Bioscience #610832), secondary antibody (BD Bioscience #554002, BD Bioscience #554021).

2.5.2 Tandem Monomeric RFP-GFP-Tagged LC3 (tfLC3) Reporter System

1. *Plasmid transfection*: tandem GFP-RFP-LC3 plasmid (addgene #21074), Lipofectamine™ 2000 Transfection Reagent (Invitrogen #11668027).
2. *Fixing solution*: Methanol (100%) or formalin (10%).
3. *Washing buffer*: PBS containing 0.05% Tween-20 (PBST).

3 Methods

3.1 Transmission Electron Microscopy (TEM) for Cell-Cultured Sample

1. Rinse the culture medium from samples with phosphate buffer or 0.1 M cacodylate buffer at same temperature as cells.
2. Fix the cells in 2.5% glutaraldehyde in 0.1 M sodium cacodylate buffer for 1.5 h, 4 °C.
3. Rinse in 0.1 M cacodylate buffer for 5–10 min.
4. Fix in 1% osmium tetroxide in 0.1 M cacodylate buffer for 1 h.
5. Rinse in 0.1 M cacodylate buffer for 5–10 min.
6. Dehydrate in graded ethanol series: 50%, 70%, 80%, and 95% for 5 min each.
7. Dehydrate with 100% ethanol three times, 10 min each.
8. Section the block into 60–100 nm ultrathin sections (*see* **Note 4**).
9. Picked up the sections on copper grids.
10. For routine analysis, ultrathin sections were stained with 2% uranyl acetate and lead citrate.
11. Electron micrographs were obtained using a transmission electron microscope [Panda et al. [6] (Fig. 1)].

3.2 Acidic Content Analysis

1. Seed the cells in a 6 well plate at a density of 2×10^5 cells/ml and allow them to grow for 24 h inside CO_2 incubator.
2. Add acridine orange (0.5 μg/ml) or MDC (0.05 mmol/L) and incubate for 15 min at 37 °C.
3. Discard the media and wash the cells with 1× PBS for three times.
4. Analyze in flow cytometer or fluorescence microscope (Fig. 2) (*see* **Note 5**).

3.3 Atg8/LC3, P62, and Other Autophagic Protein Detection and Quantification by Western Blot

3.3.1 Lysate Preparation (Timing up to 45 min)

1. Culture the adherent cells in 60 mm disc with desired treatment for appropriate time points (*see* **Note 8**).
2. Scrap the cells after appropriate time points in ice-cold condition and transfer the cells to the microcentrifuge tube gently.
3. Centrifuge the cells at 5000 × *g* for about 5 min at 4 °C.
4. Remove the supernatant and resuspend the pellet in 1× PBS ice-cold PBS.

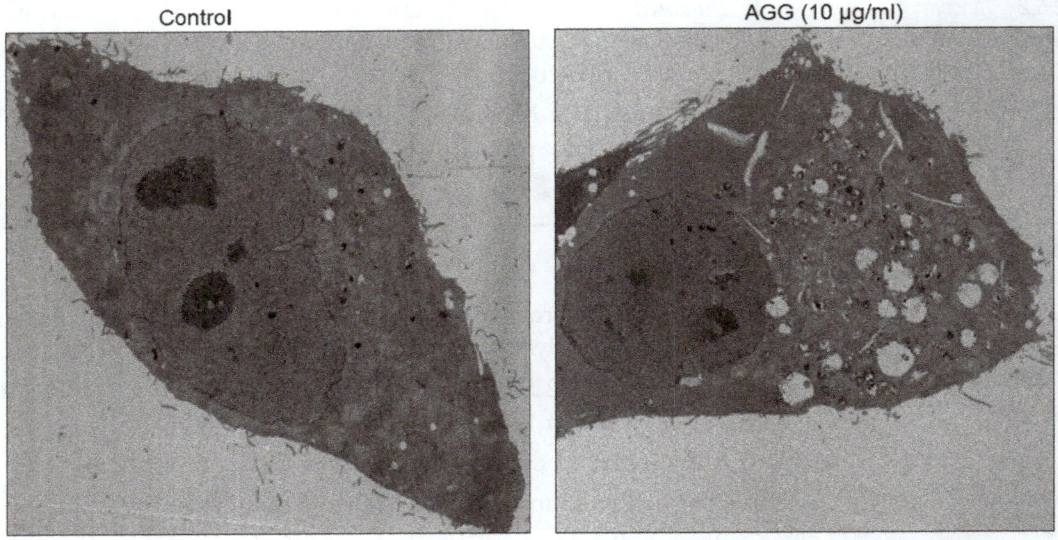

Fig. 1 HeLa cells were treated with *Abrus* agglutinin (10 μg/ml) for 24 h, and cells were fixed and processed for electron microscopy (from Panda et al., Mol Carcinog. 2017;56(2):389–40)

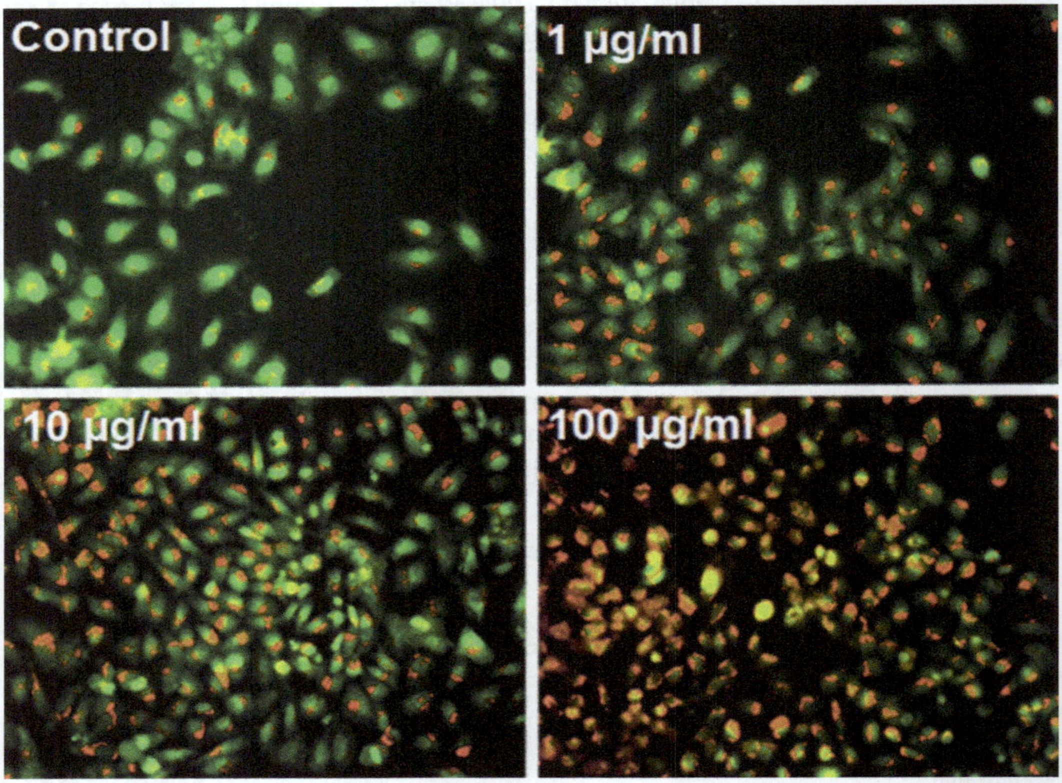

Fig. 2 HeLa cells were treated with different concentrations (1, 10, and 100 μg/ml) of peanut agglutinin for 24 h followed by staining with acridine orange and then visualized in inverted fluorescence microscope (Olympus IX71; 200) (from Mukhopadhyay et al., Food Chem Toxicol. 2014;64:369–77)

5. Centrifuge the cells at 5000 × *g* for about 5 min at 4 °C.
6. Remove the supernatant and apply freshly prepared 1× cell lysis buffer.
7. Lyse the cells by applying 1× cell lysis buffer mixed with protease and phosphatase inhibitor.
8. Incubate the cells for 30 min at 4 °C for complete lysis in nondenaturing condition.
9. Centrifuge cells at 14,000 × *g* for 7 min at 4 °C in a cooling centrifuge and collect the supernatant (*see* **Note 9**).

3.3.2 Estimate the Protein of Supernatant by Bradford Method (Timing up to 30 min)

1. Prepare the Bradford reagent (BR) by diluting it five times with deionized water (DI H_2O) (one part BR and four parts DI H_2O).
2. Mix 2 μl of cell lysate with working solution of BR.
3. Take the absorbance at 595 nm.
4. Estimate the unknown protein concentration using BSA standard plot.
5. Load about 20–100 μg of proteins for band detection (protein quantity depends upon the cells as well as protein).

3.3.3 Sodium Dodecyl Sulfate Polyacrylamide Gel Electrophoresis (SDS-PAGE) (Timing up to 2 h)

1. Prepare the samples by adding 1× Laemmli sample buffer.
2. Boil the samples at 100 °C for 5 min to denature (*see* **Note 10**).
3. Centrifuge the samples for 2 min at 14,000 × *g* for proper mixing.
4. Load equal amount of proteins in each well (keep equal volume in each well).
5. Load 5 μl of pre-stained protein ladder as standard molecular weight.
6. Run the gel at 100 V until the dye reach the bottom.

3.3.4 Protein Transfer and Blocking (Timing up to 2 h 40 min)

7. Transfer protein to the nitrocellulose or PVDF membrane at 4 °C for 100 min at 100 V.
8. Block the membrane with blocking buffer (5% BSA/skim milk prepared in 0.1% PBST) for 1 h at room temperature or overnight at 4 °C.

3.3.5 Antibody Incubation (Timing up to 16 h)

9. Incubate the membrane with primary antibody for overnight (*or 12 h*) on a shaking platform (shaker) at 4 °C.
10. Wash the membrane thrice in PBST (10 min each).
11. Incubate membrane in secondary antibody for 2 h at room temperature.
12. Wash the membrane in PBST thoroughly for 30 min (3 × 10 min) to avoid unnecessary background noise.

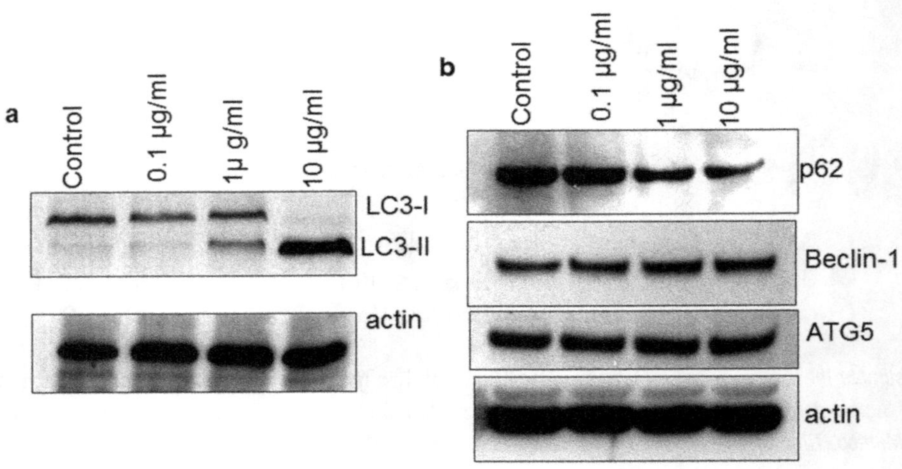

Fig. 3 After 24 h of *Abrus* agglutinin (AGG) treatment, LC3-II expression was analyzed by Western blot (**a**). HeLa cells were treated with AGG (10 μg/ml) for 24 h, and expression of p62, Beclin-1, and ATG5 were analyzed by Western blot (**b**) (from Panda et al., Mol Carcinog. 2017; 56(2):389–40)

13. Treat the blot with chemiluminescence substrate for detection as well as documentation of the protein band (Fig. 3a, b) using Image Quant LAS500 (GE Healthcare, USA).
14. Measure the band intensity by densitometry applying ImageJ software (*see* **Note 11**).

3.4 GFP-Atg8/LC3 Fluorescence Microscopy

GFP-LC3 provides a marker that is relatively easy to use for monitoring autophagy induction (based on the appearance of puncta) through direct fluorescence microscopy or indirect immunofluorescence.

3.4.1 Seeding and Transfection (Timing up to 15 h 30 min)

1. Seed cells to be 70–90% confluent at transfection.
2. Remove the complete media and acclimatize the cells with Opti-MEM® Medium for 30 min.
3. Dilute referred amounts of Lipofectamine ™ 2000 Reagent in Opti-MEM® Medium and 10–20 μg of GFP-LC3 plasmid in Opti-MEM® Medium, and incubate each for 5 min at RT.
4. Add diluted GFP-LC3 plasmid to diluted Lipofectamine® 2000 Reagent (1:1 ratio).
5. Incubate the mixture for 20 min at RT.
6. Add DNA-lipid complex to cells and incubate for 3 h.
7. Replace the media with complete media and incubate at 37 °C in CO_2 incubator for 12 h.

3.4.2 Double Transfection (Timing up to 15 h 30 min)

8. Repeat the transfection with the protein of interest by following **steps 2–7** (*see* **Note 12**).
9. Incubate the cells for desired compound with appropriate time point (*see* **Note 13**).

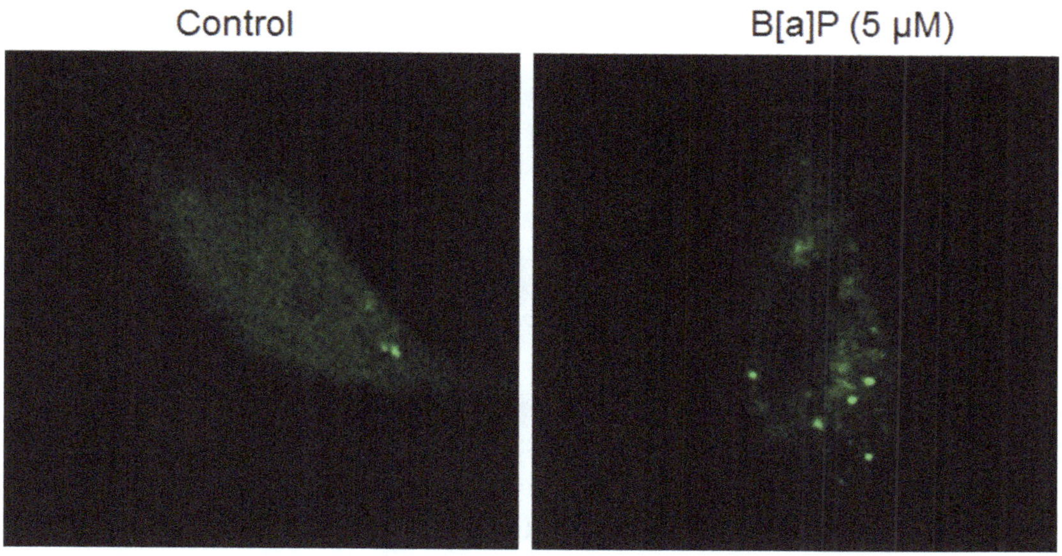

Fig. 4 HaCaT cells were transfected with GFP-LC3 plasmid for 48 h and treated with benzo[a]pyrene (5 μM) for 6 h; green fluorescence puncta for autophagosome formation were examined by confocal microscopy (from Das et al., Free Radic Biol Med. 2017; 112:452–463)

3.4.3 Visualization/ Analysis of Transfected Cells (Timing up to 45 min)

10. After desired time point, remove the media.
11. Fix the cells with chilled methanol (100%) at −20 °C for 20 min or with formalin (10%) at room temperature for 10–20 min.
12. Remove the methanol and wash it with 1× PBS for 30 min (3 × 10 min each).
13. Add DAPI to each plate/well to stain nucleus.
14. Observe the cells under fluorescence microscope and count the green puncta (Fig. 4) (*see* **Note 14**).
15. Quantify the puncta level per cell by manual counting and colocalization of GFP-LC3-II with protein of interest using ImageJ (*see* **Note 15**).

3.5 Measurement of Autophagic Flux

Autophagic flux is often inferred on the basis of LC3-II turnover, measured by Western blot in the presence and absence of lysosomal or vacuolar degradation.

3.5.1 LC3-II/Atg8–PE Turnover

1. Culture the adherent cells in 60 mm disc with desired treatment for appropriate time points (*see* **Notes 16** and **17**).
2. Scrap the cells after appropriate time points in ice-cold condition, and transfer the cells to the microcentrifuge tube gently.
3. Centrifuge the cells at 5000 × *g* for about 5 min at 4 °C.
4. Follow **step 4** onward of Western blotting lysate preparation (Fig. 5a).

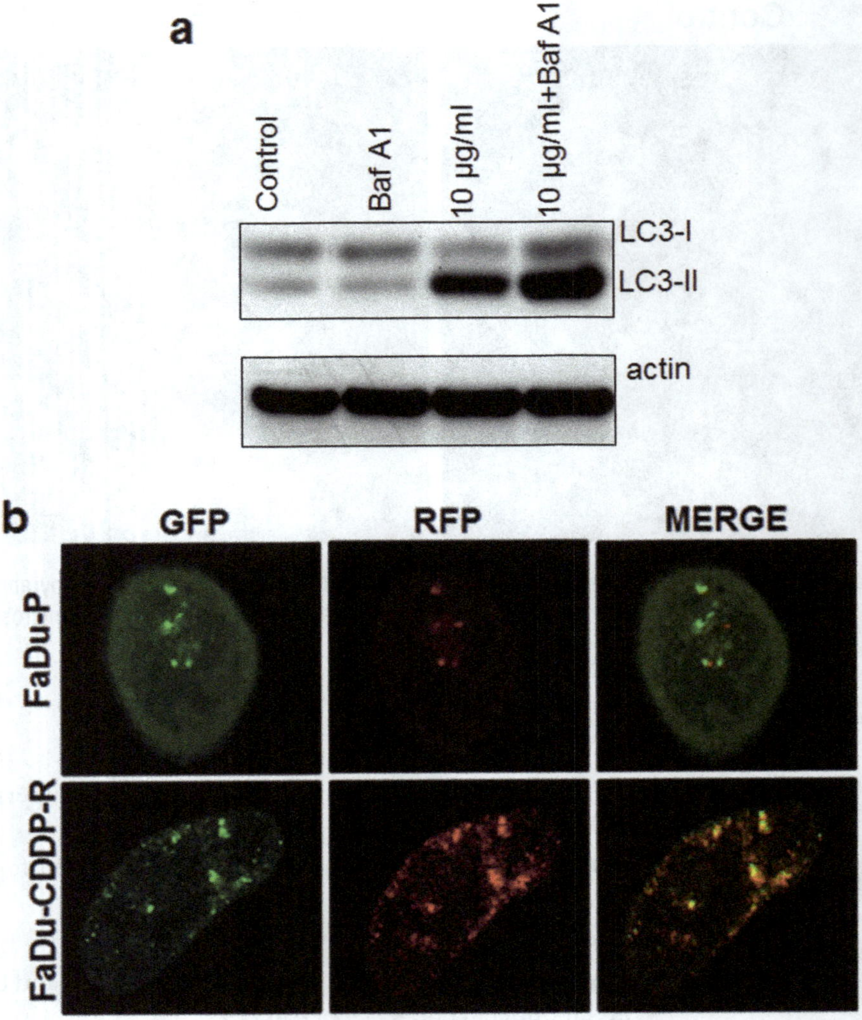

Fig. 5 HeLa cell was treated with *Abrus* agglutinin (10 μg/ml) in the presence of bafilomycin A1 (100 nM) in 24 h, and LC3-II expression was analyzed to demonstrate autophagic flux (**a**) (from Panda et al., Mol Carcinog. 2017; 56(2):389–40). Further, autophagic flux analysis in parental and resistant FaDu cells was performed after tF-LC3 transfection, and the numbers of RFP^{+} GFP^{+} (yellow) and RFP^{+} GFP^{-} (red) puncta per cell were representing autophagosome and autolysosome, respectively, and were counted (**b**) [From Naik et al., Cell Prolif. 2018;51(1)]

3.5.2 Tandem Monomeric RFP-GFP-Tagged LC3 (tfLC3) Reporter System

In this section we discuss about a fluorescence assay designed for autophagic flux measurement called traffic light assay, which relies on the use of a tandem monomeric RFP-GFP-tagged LC3 (tfLC3).

Seeding and Transfection (*Timing* up to 15 h 30 min)

1. Seed cells to be 70–90% confluent at transfection.
2. Dilute four amounts of Lipofectamine® Reagent in Opti-MEM® Medium.

3. Dilute tf-LC3 plasmid in Opti-MEM® Medium.
4. Add diluted tf-LC3 plasmid to diluted Lipofectamine® 2000 Reagent (1:1 ratio).
5. Incubate for 5 min at RT.
6. Add DNA-lipid complex to cells and incubate for 3 h.
7. Replace the media with freshly prepared media and incubate at 37 °C inside CO_2 incubator for 12 h.
8. Incubate the cells for desired compound with appropriate time point.

Visualization/Analysis of Transfected Cells (*Timing* up to 45 min)

9. After the desired time point, remove the media.
10. Fix the cells with chilled methanol (100%) at −20 °C for 20 min.
11. Remove the methanol and wash it with 1× PBS for 30 min (3 × 10 min each).
12. Add DAPI to each plate to stain nucleus.
13. Observe the cells under fluorescence microscope and count the red, green, and yellow puncta (Fig. 5b).
14. Quantify the puncta level using ImageJ (*see* **Note 18**).

4 Notes

1. Glutaraldehyde, formaldehyde, and paraformaldehyde are irritants to the eyes, skin, and respiratory tract.
2. If using the crystalline form of osmium tetroxide, this solution must be prepared a day before usage, as it is poorly soluble in water. Osmium tetroxide is highly toxic and must be used under a fume hood. It stains the skin black and its vapors can cause temporary blindness.
3. Uranyl salts dissolve slowly in water and are light sensitive and toxic if ingested, inhaled, and absorbed through a skin defect; effects are cumulative. Use under a fume hood.
4. Use fume hood while working with lead citrate. The pH should be 12.0 ± 0.1; if not, a precipitate will form while staining grids. If the solution is cloudy after adding NaOH, discard and make a fresh solution. To prevent sections from wrinkling, ensure the temperature of the hot plate is at least 70 °C.
5. Acridine orange does not emit green fluorescence in an acidic compartment. Whereas, MDC is extremely light sensitive. So incubation is needed to be done in the dark, and the data should be collected without any delay after staining without any exposure to intense light.

6. Tris can be dissolved faster provided the water is warmed to about 37 °C. However, the downside is that care should be taken to bring the solution to room temperature before adjusting pH. Unpolymerized acrylamide is a neurotoxin and care should be exercised to avoid skin contact. The acrylamide solution can be stored at 4 °C for 1 month.
7. SDS precipitates at 4 °C. Therefore, the lysis buffer needs to be warmed prior to use.
8. When dealing with animal tissues, Western blotting of LC3 should be performed on frozen biopsy samples homogenized in the presence of protease inhibitors.
9. Cell lysate can be stored for up to 1 month at −20 °C.
10. LC3-I is more labile than LC3-II, being more sensitive to freezing-thawing and to degradation in SDS sample buffer. Therefore, fresh samples should be boiled and assessed as soon as possible and should not be subjected to repeated freeze-thaw cycles.
11. In mammalian cells, however, the total levels of LC3 do not necessarily change in a predictable manner, as there may be increase in the conversion of LC3-I to LC3-II or a decrease in LC3-II relative to LC3-I if degradation of LC3-II via lysosomal turnover is particularly rapid. Both of these events can be seen sequentially in several cell types as a response to total nutrient and serum starvation.
12. The time for transient expression of GFP-LC3 varies from 24 to 48 h depending upon cell types, e.g., HeLa cells may require 1 day, whereas neuronal cell lines such as SH-SY5Y cells need at least 48 h.
13. Preincubation of cells expressing GFP-LC3 with leupeptin stabilizes the GFP-LC3 signal during fluorescence microscopy by inhibiting the degradation of membrane-conjugated GFP-LC3 present within autolysosomes under induced autophagic flux.
14. Colocalization of LC3-II along with additional marker like ATG12-ATG5-ATG16L1 specifies the identity of an LC3-positive structure (autophagosome).
15. Some treatments lead to the "rounding up" of cells resulting in apparent changes in the number of GFP-LC3 puncta per cell. Therefore, to avoid misinterpretation of results, manual review of cell images is highly recommended. If cells are rounding up due to apoptosis or mitosis, it is easy to automatically remove them from analysis based on nuclear morphology (using DAPI or Hoechst staining) or cell roundness.

16. When using autophagy inhibitors, it is of fundamental importance to assess proper conditions of inhibitor concentration and time of preincubation to ensure full inhibition, such as chloroquine (1–4 h), bafilomycin A1 (1–4 h), pepstatin A (>8 h), leupeptin (6 h preincubation), and E-64d (1 h of preincubation).
17. The relevant parameter in LC3 assays is the difference in the amount of LC3-II in the presence and absence of saturating levels of inhibitors, which can be used to examine the transit of LC3-II through the autophagic pathway; if flux is occurring, the amount of LC3-II will be higher in the presence of the inhibitor.
18. *Tandem fluorescent markers show real-time changes in autophagosome fusion with lysosomes;* the GFP signal is sensitive to the acidic and/or proteolytic conditions of the lysosome lumen, whereas mRFP is more stable. Therefore, colocalization of both GFP and mRFP fluorescence indicates a compartment that has not fused with a lysosome, such as the phagophore or an autophagosome. In contrast, an mRFP signal without GFP corresponds to an amphisome or autolysosome.

Acknowledgments

Research support was partly provided by the Department of Biotechnology [Grant Number: BT/PR7791/BRB/10/1187/2013], Board of Research in Nuclear Sciences (BRNS) [Number: 37(1)/14/38/2016-BRNS/37276], Department of Atomic Energy (DAE), and Science and Engineering Research Board (SERB) [Number: EMR/2016/001246]. Research infrastructure was partly provided by Fund for Improvement of S&T Infrastructure in Universities and Higher Educational Institutions (FIST) [Number: SR/FST/LSI-025/2014], Department of Science and Technology, Government of India.

References

1. Bhutia SK, Mukhopadhyay S, Sinha N, Das DN, Panda PK, Patra SK, Maiti TK, Mandal M, Dent P, Wang XY, Das SK, Sarkar D, Fisher PB (2013) Autophagy cancer's friend or foe? Adv Cancer Res 118:61–95
2. Panda PK, Mukhopadhyay S, Das DN, Sinha N, Naik PP, Bhutia SK (2015) Mechanism of autophagic regulation in carcinogenesis and cancer therapeutics. Semin Cell Dev Biol 39:43–55
3. Mukhopadhyay S, Panda PK, Sinha N, Das DN, Bhutia SK (2014) Autophagy and apoptosis: where do they meet? Apoptosis 19(4):555–566
4. Panda PK, Naik PP, Meher BR, Das DN, Mukhopadhyay S, Praharaj PP, Maiti TK, Bhutia SK (2018) PUMA dependent mitophagy by Abrus agglutinin contributes to apoptosis through ceramide generation. Biochim Biophys Acta 1865(3):480–495
5. Mukhopadhyay S, Naik PP, Panda PK, Sinha N, Das DN, Bhutia SK (2016) Serum starvation induces anti-apoptotic cIAP1 to

promote mitophagy through ubiquitination. Biochem Biophys Res Commun 479 (4):940–946

6. Panda PK, Behera B, Meher BR, Das DN, Mukhopadhyay S, Sinha N, Naik PP, Roy B, Das J, Paul S, Maiti TK, Agarwal R, Bhutia SK (2017) Abrus Agglutinin, a type II ribosome inactivating protein inhibits Akt/PH domain to induce endoplasmic reticulum stress mediated autophagy-dependent cell death. Mol Carcinog 56(2):389–401
7. Mukhopadhyay S, Schlaepfer IR, Bergman BC, Panda PK, Praharaj PP, Naik PP, Agarwal R, Bhutia SK (2017) ATG14 facilitated lipophagy in cancer cells induce ER stress mediated mitoptosis through a ROS dependent pathway. Free Radic Biol Med 104:199–213
8. Mukhopadhyay S, Sinha N, Das DN, Panda PK, Naik PP, Bhutia SK (2016) Clinical relevance of autophagic therapy in cancer: investigating the current trends, challenges, and future prospects. Crit Rev Clin Lab Sci 53 (4):228–252
9. Das DN, Naik PP, Mukhopadhyay S, Panda PK, Sinha N, Meher BR, Bhutia SK (2017) Elimination of dysfunctional mitochondria through mitophagy suppresses benzo[a]pyrene-induced apoptosis. Free Radic Biol Med 112:452–463
10. Mukhopadhyay S, Das DN, Panda PK, Sinha N, Naik PP, Bissoyi A, Pramanik K, Bhutia SK (2015) Autophagy protein Ulk1 promotes mitochondrial apoptosis through reactive oxygen species. Free Radic Biol Med 89:311–321
11. Mukhopadhyay S, Panda PK, Behera B, Das CK, Hassan MK, Das DN, Sinha N, Bissoyi A, Pramanik K, Maiti TK, Bhutia SK (2014) In vitro and in vivo antitumor effects of Peanut agglutinin through induction of apoptotic and autophagic cell death. Food Chem Toxicol 64:369–377
12. Naik PP, Mukhopadhyay S, Panda PK, Sinha N, Das CK, Mishra R, Patil S, Bhutia SK (2018) Autophagy regulates cisplatin-induced stemness and chemoresistance via the upregulation of CD44, ABCB1 and ADAM17 in oral squamous cell carcinoma. Cell Prolif 51 (1). https://doi.org/10.1111/cpr.12411

Methods in Molecular Biology (2018) 1854: 223–228
DOI 10.1007/978-1-4939-8748-1

INDEX

A

B

C

D

E

F

G

H

I

L

M

N

O

P

Q

R

S

CPSIA information can be obtained
at www.ICGtesting.com
Printed in the USA
LVHW06*1801181018
594044LV00002B/32/P

9 781493 987474